PRAISE FOR *CELSIUS: A LIFE AND DEATH BY DEGREES*

'From my home in Canada, where 'fire' is now a season, I read *Celsius* to learn about the man whose name signals both threat and comfort. What I found was a life fuelled by insatiable curiosity and an ability to wonder – much needed qualities as we face today's polycrisis. This biography reveals the human capacity to seek answers, even when, as Celsius writes: 'the process of discovery has no end'. This is an important book for our time.'

KATE ELLIOTT
Simon Fraser University
Vancouver

'This book gives readers a much broader perspective on the life of Anders Celsius and his diverse scientific work and discoveries. The story takes us deep into the world where his experimental measurements in Lapland with Pierre Louis Moreau de Maupertuis showed how Newton's universal law of gravitation literally shapes our planet. It was a breakthrough that forged the later lives and fortunes of both men, and forever changed the way humanity perceives its home.'

Veli-Markku Korteniemi
Chairman of the Maupertuis Foundation
Finland

'This is a beautifully written masterwork of science biography, covering the life and times of the great Swedish astronomer Anders Celsius. Ian Hembrow provides a gripping narrative that details his search for a man who was very much embedded in his city and landscape, who now remains relatively obscure even though he left an indelible mark above and beyond his invention of the Celsius temperature scale. This biography not only sheds light on a brilliant polymath and the colorful world of eighteenth-century science and its many characters; it also raises larger and more contemporary questions around global warming and the relevance of Celsius to the world we live in today.'

Dr Sarah Covington
Professor of History, Biography and Memoir
City University of New York

CELSIUS
A LIFE AND DEATH BY DEGREES

IAN HEMBROW

Front cover image: Etching of Anders Celsius *c.* 1730 by an unknown artist, and Ed Hawkins' warming stripes, University of Reading, 2024.
Back cover image: Portrait of Anders Celsius *c.* 1730 by Olof Arenius.

First published 2024

The History Press
97 St George's Place, Cheltenham,
Gloucestershire, GL50 3QB
www.thehistorypress.co.uk

© Ian Hembrow, 2024

The right of Ian Hembrow to be identified as the Author of this work has been asserted in accordance with the Copyright, Designs and Patents Act 1988.

All rights reserved. No part of this book may be reprinted or reproduced or utilised in any form or by any electronic, mechanical or other means, now known or hereafter invented, including photocopying and recording, or in any information storage or retrieval system, without the permission in writing from the Publishers.

British Library Cataloguing in Publication Data.
A catalogue record for this book is available from the British Library.

ISBN 978 1 80399 461 1

Typesetting and origination by The History Press.
Printed and bound in Great Britain by TJ Books Limited, Padstow, Cornwall

 Trees for LYfe

This book is dedicated to Martin Ekman, associate lecturer at Uppsala University – a friend and scientist who carries his knowledge and wisdom lightly, and who has done so much to keep the name, achievements and personality of Anders Celsius alive for the twenty-first century.

It is easy for us to stand outside, our face to the wind, and not realize that once upon a time we had no name, nor understanding of what wind was.[1]
Carl Leonard

1.5°C is not a goal or target, it is a biophysical limit. Cross it, and we are likely to trigger multiple tipping points in the Earth system. There is no safe landing for humanity, with regards to a manageable climate, unless we also bend the curves and return back to a safe operating space within planetary boundaries, for land, water, biogeochemical flows and biodiversity.[2]
Professor Dr Johan Rockström

Three things cannot be long hidden: the sun, the moon and the truth.[3]
Attrib. Buddha (c. 563–483 BCE)

CONTENTS

Foreword	9
Prologue	11
Charts and Maps	17
Anders Celsius' Timeline	23

PART I: FIRE

1 Risen from the Sea: The Land from Which Celsius Came and Where he Rests Among Kings — 27

2 Forged in Flames: The Origin Story of Celsius the Scientist (1701–02) — 37

3 A City of Learning: Uppsala and Sweden's Age of Freedom (1720–72) — 45

4 A Family of Ambition: The Celsius Name and Prominent Ancestors — 51

5 A Frustrated Father: Nils Celsius and his Ruinous Dispute with the Church of Sweden — 57

6 The Age of Enlightenment: Young Celsius and Science in Eighteenth-Century Europe (1710–19) — 65

PART II: LIGHT AND AIR

7 The Young Professor: Establishing his Roles and Reputation in Uppsala (1719–32) — 81

8 Celsius' Grand Tour: Learning from Europe's Great Astronomers and Observatories — 99

9 Two Countries, Four Sisters: Travels and Studies in Germany and Italy (1732–34) — 105

| 10 | In Paris: Encounters, Observations and Opportunity (1734–35) | 119 |
| 11 | In London: Appreciation, Trust, Recognition and Craftsmanship (1735–36) | 133 |

Part III: Land

12	Towards the Pole: The French King's Arctic Expedition Led by de Maupertuis (1736–37)	149
13	Arctic Summer: The Torne Valley, Instruments and Triangulation	161
14	Arctic Winter: Ice, Measurement and Calculation	177
15	Fighting for the Truth: Conflict and Controversy (1737–40)	185

Part IV: Sea and Space

16	Serving and Observing: Uppsala's First Observatory (1741)	207
17	A Vast, Profound Truth: Unravelling the Baltic Sea Mystery (1741–43)	215
18	The Infinite and the Invisible: Magnetism and Making Connections (1740–43)	231

Part V: Temperature and Climate

19	One Hundred Steps: Creating the Centigrade Scale (1741–43)	239
20	Death of a Star: Celsius' Illness and Death (1743–44)	251
21	Noble Successors: How Wargentin, Hiorter, Strömer and Others Continued Celsius' Work	257

Part VI: Time

| 22 | A Safe and Just Earth for All: Celsius' Legacy, Our Indifferent Planet and Hope | 271 |

Epilogue	281
Bibliography	285
Acknowledgements	287
Notes	291
Index	299

FOREWORD

By Anna Rutgersson, Professor of Meteorology,
University of Uppsala, Sweden

In 1722, 20-year-old Anders Celsius and his professor Eric Burman began making temperature observations at the university in the old centre of Uppsala. These still continue today at my department nearby.

Three hundred years after these first measurements, we held a celebration conference to mark the tricentenary of this important milestone in science. During the event, I tried to picture what weather and climate observations would be like in another 300 years' time. But my imagination failed.

As professor of meteorology at Uppsala University, I've built my career to a large extent on understanding the Earth's atmosphere and oceans. My research into air-water interaction and exchange, atmospheric turbulence, heat, humidity and greenhouse gases relies on systems and methods that test the limits of what it's possible to do with high-quality data in the field. As such, I feel connected to my famous predecessor Anders Celsius.

Some of the things he did and explored are obvious to us now, some were probably wrong, and others left us with new knowledge that continues to shape our modern lives. This is why he and his work are still remembered, and why he remains so important.

Celsius was driven by a wish to understand and describe the world that surrounded him. It's a curiosity that I try to keep in my own line of work and hope to inspire my students with. Celsius was well liked and very humble, but 300 years later, we still use his name every day.

My first contact with Ian, as author of this book, was around the time of the 'Celsius 300' celebration in 2022. He later spent time with my colleague, Professor Tom Stevens, and others exploring the city, story and legacy of Anders Celsius. I was impressed by Ian's enthusiasm to discover more about Celsius the man, the life he lived and how the city of Uppsala formed his character.

There are a few other books and sources of information about Anders Celsius and his work, but the engaging way his relevance for science and society today are brought to life in these pages is truly fascinating. We get to follow Ian's footsteps as he traces those of Anders Celsius, helping us to know the book's subject as a person and as a scientist, whose busy career gives us a window into life in Uppsala and beyond in the first half of the eighteenth century. We learn that there was much more to Celsius than the name he gave to the scale we use when measuring temperature.

Celsius gives us a perspective on the great challenges of today, when we more than ever realise the need to understand our surroundings and changes to the global climate and environment from the perspective of human-induced impact. We cannot prepare effectively for the future if we do not seek to observe and understand the past and present. This book helps readers to do that, be they scientists, historians or anyone with an interest in the world around them and their place in it.

I echo Anders Celsius' wish that humanity's curiosity, discovery and science can be focused upon fulfilling our brief destiny, so that we may pass on this wonderful Earth alive and untrammelled to those who follow us.

Anna Rutgersson
Geocentrum, Uppsala,
Spring 2024

PROLOGUE

IN SEARCH OF TRUTH

'That's where Professor Celsius worked,' said my companion Ralph, pointing to a squat, old, yellow and white building in the busy pedestrianised shopping street.

'What, Celsius ... as in centigrade?' I asked, halting in my stride.

'Yes.'

'Wow!'

This is how my search for Celsius and fascination with his influence began. It was my first visit to the pretty Swedish university city of Uppsala. I'd been commissioned to write a book about the World Health Organization's Collaborating Centre for Drug Monitoring there, and my host, the Englishman Professor Ralph Edwards, had taken me out for a lunchtime stroll. We walked up a steep path to the salmon pink castle, the high walls and twin domes of which dominate the city's skyline. From there, we descended to crunch across the gravel of the flawless botanical garden dedicated to Uppsala's famous scientific son, the taxonomist Carl Linnaeus (1707–78), and then wound our way back into the modern city centre, where we now stood in the pedestrianised shopping street of Svartbäcksgatan.

The building in front of us, which had brought me to such a sudden, standing stop was Scandinavia's first purpose-built astronomical observatory, created here in 1741 by another of the city's and country's most notable Enlightenment figures, Anders Celsius. For some reason, the mention of this person's name, in that place, at that moment, sent a tingle through my whole body. I felt connected to something urgent and important.

So this was where Celsius, the man whose name – now enshrined in the internationally agreed targets to tackle global climate change, which frame the whole future of humanity[1] – went about his business. This was where he scanned the skies, observed the stars and invented his eponymous temperature scale. I knew that the painfully negotiated United Nations agreements were all about limiting future average temperature rises to just a few degrees Celsius above pre-industrial levels. But what else did I know about this familiar name and the scientist behind it? Like most people, I realised, very little. But, in that instant, I was seized by a desire to discover more.

I didn't have to wait long. A short distance along the street we encountered a more-than-life-size bronze statue of Anders Celsius. For a man who lived in the first half of the eighteenth century, it's a curiously modern and figurative depiction. He stands slim and erect, gazing to the heavens with a sextant raised in his outstretched left hand and a long bulb thermometer in his right. The stylised tails of a periwig and long frock coat splay out behind him, and he perches on top of a tapering, tiered arrangement of spouts and spheres that once cascaded with water, but then lay dusty and dry. I wondered if this was an ominous, if unintentional, visual metaphor for the state of our planet. At the bottom of the sculpture is a beachball-sized Earth globe – its key meridians and lines of latitude cast in the dark metal.

Apart from a perky, upturned nose, the statue has no facial features and few other surface details. But it immediately suggests a sprightly, confident and gracious young man – an enquiring mind and energetic spirit. Unlike many statues, this figure seemed to have life and personality, and it set me thinking: what could this long-dead person tell me about my life today, and the future prospects for humanity? What should we be aware of every time we use his name? And what might we learn from him, his work and the time in which he lived?

This book tells how Anders Celsius progressed from child to man, and from bright student to a fully fledged scientific great, whose career was cut tragically short. It examines who and what he became, and how the city and landscape where he spent most of his life shaped him. It also traces my own journey to follow in his footsteps and establish his lasting relevance. It seeks to rescue Celsius the man from obscurity and restore him to his rightful place among the best-known names of science and history.

In the months that followed, and on subsequent visits to Uppsala and other parts of Sweden, I learned more about Celsius. I discovered that he came from an illustrious family line of Swedish astronomers and mathematicians, and that he died from tuberculosis in 1744 when he was just 42. I also found out that his achievements extended far beyond inventing the temperature scale that defines most people's knowledge. In fact, his work on thermometry came right at the end of his life, a footnote to a far broader career.

Anders Celsius was, I rapidly came to recognise, a theorist and practitioner of extraordinary range and long-term vision: a master of not just astronomy, but the whole realm of natural philosophy – mathematics, geophysics, geodesy – and a multilingual pioneer of measurement, data and analysis. Celsius was arguably the father of all climatology and Earth sciences. His life and work stand at the confluence of an astounding rollcall of the most influential European thinkers and leaders – from French and Swedish royalty to Pope Clement XII, Voltaire, Descartes, Newton and the Cassinis. Among scientific giants, his star and the constellation of his collaborators shone particularly bright.

As the layers of this mercurial man and his life were revealed to me, I began to wonder what Celsius would make of our species in the twenty-first century. What might he think about humanity's slow awakening to its destructive habits and the overdue efforts to reduce or reverse our impact upon the fleck of space we call home? I imagined that he would see this not as a problem of science or nature, but as a problem of humankind, and one that only we can resolve.

By telling Celsius' story I've sought to explore these questions and suggest answers to some of them. At the dawn of the Anthropocene Age, the effects of human-induced global warming are fast outstripping the planet's natural ability to heal itself. And the *Homo sapiens* species stands incontrovertibly answerable for an accelerating mass extinction of thousands of other organisms. The United Nations estimates that around 150 species now become extinct every day.[2] Against this backdrop, what can, should or must we learn from the curiously forgotten Anders Celsius?

From my research into his life and from following his path across Europe's cities and into the frozen beauty beyond the Arctic Circle, I'm convinced that the quiet Uppsala scholar still has much to teach us. These lessons come not just from his trailblazing methods and breakthrough discoveries, but also his patient manner, his courtesy and deep-rooted instincts for international collaboration and long-term improvement.

In an academic career lasting barely two decades, Celsius was able to peer through the narrow aperture of his present to see and understand both past and future aeons in their true context. He recognised the universal forces that shape our world, and grasped the opportunities and obligations that rest with humans during our fleeting presence. Like his statue in Uppsala's Svartbäcksgatan, he stood on top of the world, and he gave his life to bequeath better knowledge to later generations. The key question for us, three centuries after his death, is: are we inclined to look, listen and take heed of these messages from the past?

The observatory building where Ralph and I stopped stands at a diagonal to the modern street, its sharp corners interrupting the line of otherwise flat, modern frontages. This is because it pre-dates the last of several fires that consumed much of the old city. Each time Uppsala burnt down and was rebuilt, the street pattern altered, reflecting perhaps the citizens' desire for a new start. So Celsius' edifice stands slightly askew – the road's only evidence of a medieval urban layout lost to merciless flames and heat.

The building's appearance suggests something of its creator too – a man who stood together yet aside from others, a modest but prominent scientist, simultaneously self-effacing and high achieving. He was able to perceive things his predecessors and contemporaries could not, and his impact still reverberates.

Celsius was an immensely *practical* scientist: an accomplished craftsman, draughtsman, technician and administrator, as well as a theorist. He was devoted to exhaustive, empirical observation and experiment in search of fundamental truths. And he embodied the new mood and style of the European Enlightenment – a quest to comprehend the elemental drivers of creation, and humanity's place within it.

Today's thinkers, strategists, decision-makers and opinion-formers face the same questions. How and where do we stand on Earth, in our solar system and the universe? If, as history and current circumstances would suggest, humans are hardwired towards expansion, dominance, conflict and consumption, how can we continue to thrive and survive within a largely sealed system of finite resources? And how are we obligated to act, now that we're aware of the damage we've already wrought and continue to inflict? If it's already too late to save ourselves, do we have any higher responsibility for the continuance and well-being of our planet? To whom or what, if at all, are we accountable?

If today's humans are as enlightened, informed and inventive as we so often assert, we will do well to look back into the life and accomplishments of Anders Celsius. Encoded within his methods, his feats, his character and the clues he left behind are some of the answers we seek. I hope that, from knowing more about this exceptional scientist, readers will gain some fresh perspective on contemporary dilemmas. And I believe that adopting a fresh outlook is a first step towards positive empowerment, agency and choices.

As the twentieth-century geochemist and guru of global warming, Professor Wallace Broecker, said: 'The Earth's climate is an angry beast, and we're poking it with sticks.'³ Humanity needs to stop goading and seeking to dominate this creature, and find peaceable ways to live with it.

Even without the significance of his work for modern-day concerns, Celsius' tale is captivating. From narrowly escaping death as a six-month-old baby in the 1702 Great Fire of Uppsala, to his four-year, shoestring-budget Grand Tour of Europe, then helping to prove the shape of the Earth through life-shortening hardship in the frozen far north, he practised science at the extremes. Celsius descended deep mines and climbed high towers to study air pressure, studied distant stars, galaxies and the Northern Lights, contributed to unravelling the mystery of apparently falling sea levels in the Baltic Sea and standardised temperature measurement using the boiling point of water and melting point of ice. By stepping into these extremes he was able to interpret and explain what lay between – the timeless actions of gravity, radiation, magnetism, tides, tectonics, vulcanism, geology, weather and waves that form our ever-changing environment.

The account presented here focuses on real people, places and events. Celsius, his peers, collaborators and successors left behind a hoard of written and physical material – not least his original 1741 thermometer and other custom-built instruments, plus an extensive library of books and correspondence. But there are gaps, so some details of Celsius' motivations, conversations and emotions are necessarily speculative, based on knowledge of his relationships, traits and the norms of the time.

I'm a storyteller, and humanity needs stories now more than ever before. The ones we discover and tell each other about history and the present can point the path ahead to our future. And to influence it in the right ways we need to imagine that future. Stories are the way we motivate ourselves and mobilise each other to effect change. As a fellow writer said to me while I was working on this book: 'All stories are true. Stories are sticky; stories are compelling; stories work.'

A world without Anders Celsius and his polymathic achievements would be less enlightened, less connected and less able to provide a sustainable future for its human and other inhabitants. It's time for us to get to know and learn from this charismatic young man.

CHARTS AND MAPS

i   Anders Celsius' Family Tree

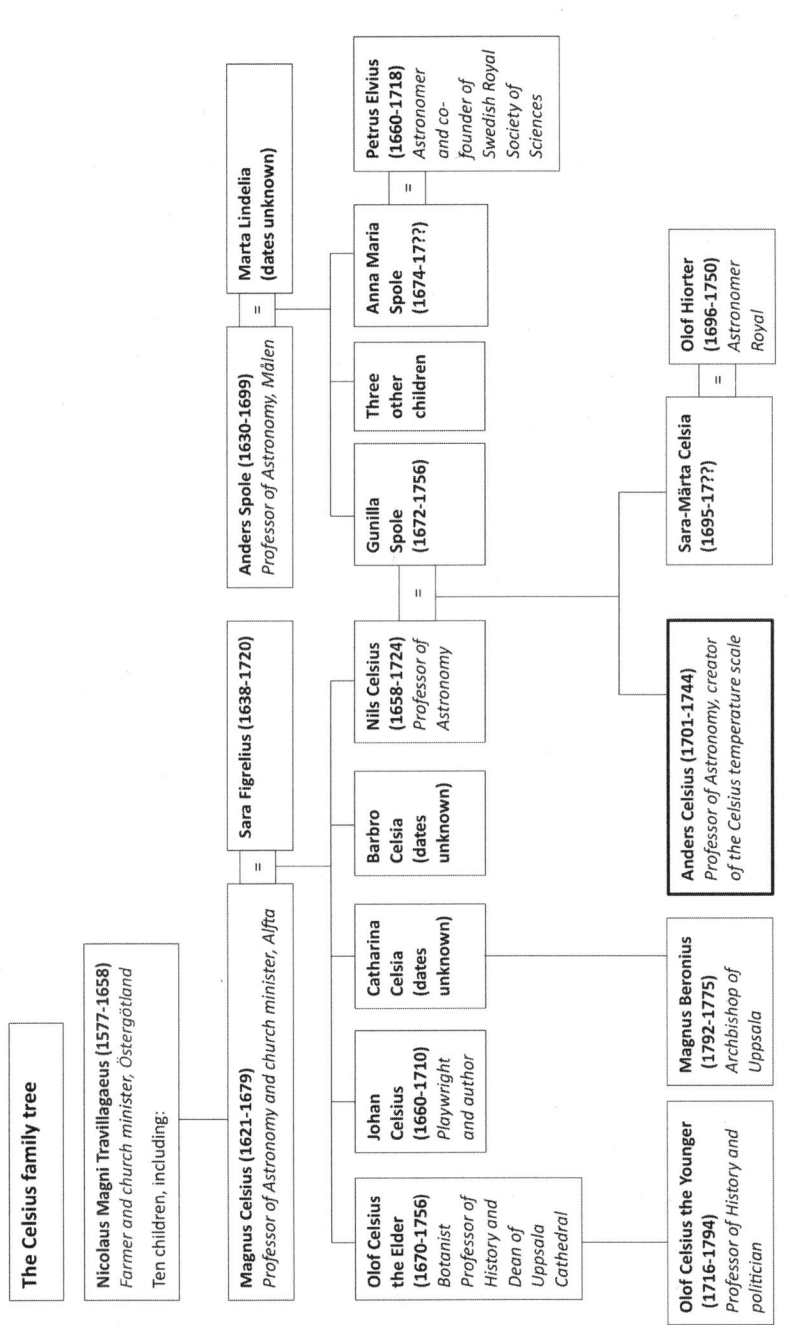

ii Celsius' Sweden (Shown on Modern Borders)

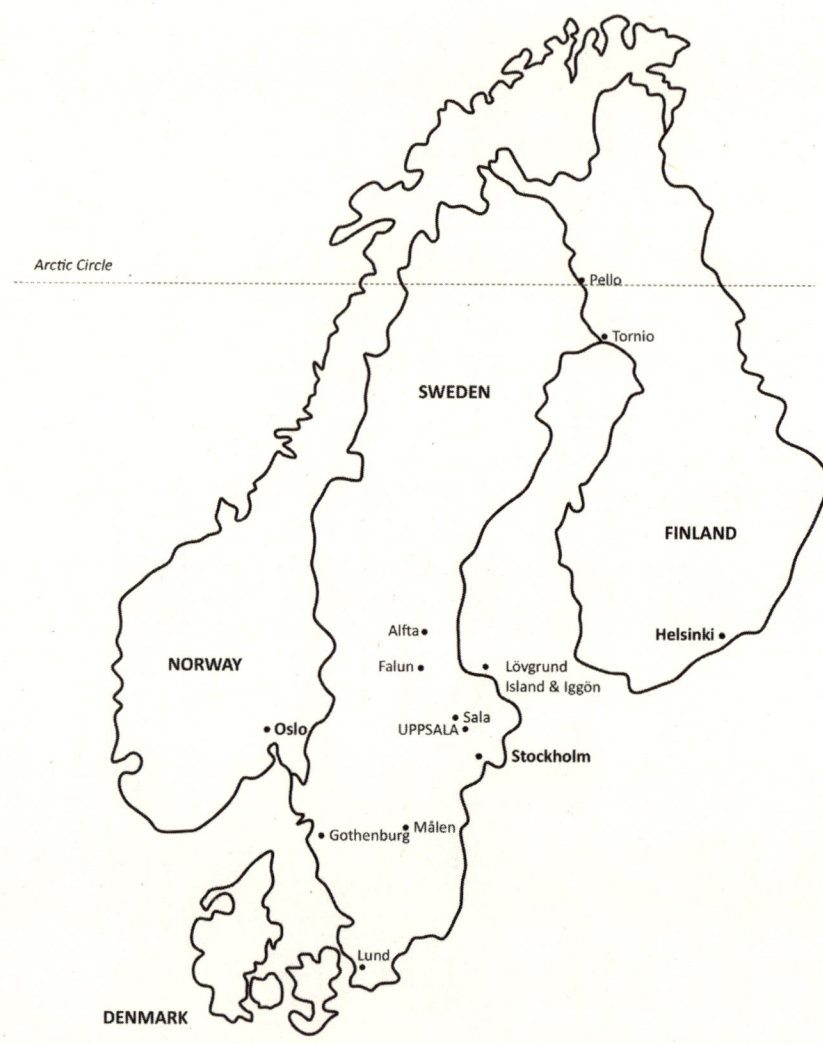

iii   Celsius' Uppsala

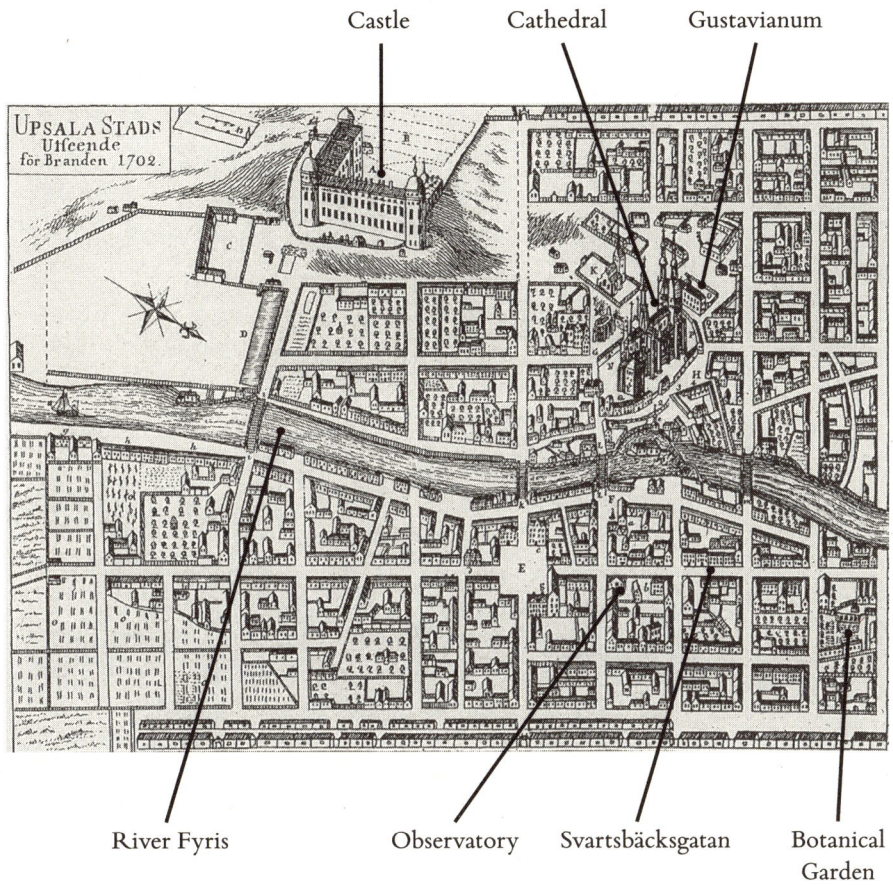

Anders Celsius' Uppsala (shown on 1702 map before the Great Fire).

iv   Celsius' European Tour 1732–37

Map created by Julie Kilburn.

ANDERS CELSIUS' TIMELINE

1701	Born in Uppsala, Sweden
1702	Family home destroyed in the Great Fire of Uppsala
1714	Began studies in law
1717	Switched to private tuition in mathematics from Anders Gabriel Duhre
1719	Began astronomy studies, taught by his father, Nils Celsius, and Eric Burman
1722	Started work as an unpaid astronomy assistant at Uppsala University
	Began systematic recording of weather data with Eric Burman
1724	Appointed as assistant to the Swedish Royal Society of Sciences
	Air pressure experiments reported by the Royal Society in London
1725	Appointed as secretary to the Swedish Royal Society of Sciences
1728	Deputised for Samuel Klingenstierna as Uppsala's professor of mathematics
1730	Appointed as professor of astronomy after Eric Burman's death
1731	Started studies of Baltic Sea levels
1732	Began scientific Grand Tour to northern Germany
1733	Continued Grand Tour through southern Germany and Italy
	Published observations of the Northern Lights

1734–35	At l'Académie des Sciences in Paris with Pierre Louis Moreau de Maupertuis
1735–36	In London, including time with Sir Edmond Halley at Greenwich Royal Observatory
1736	Returned to France to join de Maupertuis' expedition to the Arctic
1737	Returned from the Arctic and awarded a lifetime pension by King Louis XV
	Resumed role as secretary to the Royal Society of Sciences
	Built a private observatory in his mother's garden in Uppsala
1739–41	Campaigned for and built Uppsala University's first astronomical observatory in Svartbäcksgatan
1740	Resumed studies of the Northern Lights and magnetism
1741	First recorded use of the Celsius temperature scale
1743	Resumed studies of Baltic Sea levels
	Appointed rector of Uppsala University
1744	Comet observations
	Died of tuberculosis and buried in the family tomb at Gamla Uppsala
1948	Degrees Celsius adopted as the International Practical Temperature Scale
2015	United Nations Climate Change Conference (COP21) Paris Agreement to hold 'the increase in the global average temperature to well below 2°C above pre-industrial levels' and pursue efforts 'to limit the temperature increase to 1.5°C above pre-industrial levels'
2023	World's hottest year on record
	United Nations Global Stocktake, key finding 4: 'Global emissions are not in line with modelled global mitigation pathways consistent with the temperature goal of the Paris Agreement, and there is a rapidly narrowing window to raise ambition and implement existing commitments in order to limit warming to 1.5°C above pre-industrial levels'

PART I

FIRE

1

RISEN FROM THE SEA

The Land from Which Celsius Came
and Where he Rests Among Kings

What lies behind us and what lies ahead of us are tiny matters compared to what lives within us.

Attrib. Ralph Waldo Emerson

His body is hidden beneath the red carpet in the knave of the small parish church at Gamla (Old) Uppsala in southern Sweden. This is where Anders Celsius lies alongside his illustrious grandfather Magnus and unlucky father Nils. The outline of the family tombstone is clearly visible, a rectangular depression in the weave. I lifted the edge of the carpet to see a corner of the slab sealing the crypt, polished like pewter by countless feet. The youngest man who lies beneath is familiar yet concealed, close but little known.

High on the wall up to my left, a black marble plaque, carved with pale-gold Latin capitals reads (when translated):

> CLEAR SENSE, HONEST WILL,
> CAREFUL WORK AND USEFUL LEARNING
> MAY HIS BONES REST IN PEACE AS SAFELY
> AS HIS REPUTATION WILL NEVER REST.

The Celsius family line ended with Anders and his premature death in 1744 from tuberculosis. He did not marry or have children, but his legacy persists. Three centuries on, the adventures, discoveries and methods of

The memorial plaque to Anders Celsius at Gamla Uppsala church in Sweden.
(Courtesy of Dr Stephen Burt)

this restless polymath still influence the rhythms of our daily lives, and his name now carries the prospects for humankind.

Gamla Uppsala is in the heart of the Uppland region, a few kilometres north of modern Uppsala. A visit here involves just a short drive through serene suburbs. All seems calm now, but this is a landscape transformed over millennia by natural forces of immense power. And all around the church are the massive and brooding grass-covered burial mounds of the *Svea* monarchs, the tribal warlords of this part of Scandinavia. Celsius was born, lived, worked and died close to here, and it was his observations on the Baltic coast to the east that helped to untangle the mystery of how and why Uppland rose from the water.

It was the sea that brought the first people to Uppland around 5000 BCE. The early settlers were sustained by a plentiful supply of fish and the serrated coastline's myriad inlets, which steadily silted up with heavy clay to provide rich pasture and fertile farming land. Today, the gravel surfaces of graves surrounding the church are artistically raked into neat lines and swirls. Their modern appearance gives little clue to the site's significance in the region's history as the home of ancient pagan kings. Although, in one corner, an eleventh-century rune stone once used as an altar now forms part of the church's outside wall. Its Viking-era carving of a horned serpent commemorates 'Sigviðr, a traveller to England' – an augury of Celsius' own journey to England 700 years later.

Long before the Vikings, in the third century CE, it is believed that a grand, gold-adorned temple stood here, dedicated to the three great Norse gods: Odin the All-Father, Thor the thunder-god and Frey, the god of fertility. Every nine years during the month of *Goi* (mid-February to mid-March) folklore tells of the temple becoming the centre of intense worship and bloody sacrifice. Two male animals – rams, goats, boars and cockerels – were slaughtered each day during nine days of feasting, with one unlucky man also put to death alongside each pair. The bodies, both animal and human, were then strung up in a shady grove next to the temple, their slow putrefaction believed to confer holy blessings upon the trees from which they hung.

The land surrounding the village was also the scene of extravagant, ceremonial ship burials. The mounds that stretch out in all directions from the little church contain the graves of powerful chieftains, many of them buried in wooden longboats. The bodies were laid out on furs and

Woodcut of the temple at Gamla Uppsala from *Historia de gentibus septentrionalibus* (1555) by Olaus Magnus, which perfectly fitted the vision of this once being the capital of Atlantis. A golden chain winds around the building, with a human sacrifice in the spring to the right.

A 1709 map of Gamla Uppsala by Truls Arvidsson, showing some of the hundreds of burial mounds that surround the church and its Celsius family tomb.

festooned with jewels, pets and provisions: everything they would need for their journey into the afterlife. And later, in the twelfth and thirteenth centuries, the Royal Mounds formed a grand temple amphitheatre for *skeid* – carousing festivals with stallion races and more mass animal sacrifices.

With so much gore and drama soaked into this land, it is perhaps not so surprising that Olof Rudbeck the Elder (Uppsala's seventeenth-century 'universal genius' of medicine, mathematics, astronomy, botany, chemistry, physics, natural history, gothic myth, art, engineering, mechanics, archaeology and architecture) speculated that this spot was once the epicentre of the lost continent of Atlantis. To prove this theory – to his own satisfaction, if few others' – Rudbeck found linguistic, narrative and topographical evidence wherever he looked. This was, he suggested, the true home of the *Hyperborean* people of Greek mythology: a sunny, temperate and divinely blessed land beyond the north wind.

Whether Rudbeck's claims in his 1675 opus *Atlantica*[1] stand up to scrutiny or not, Celsius and his ancestors undeniably rest at a unique meeting point of history, myth and legend. The breeze that stirs the trees surrounding Gamla Uppsala church still seems to carry whispered echoes of the human struggle to claim this landscape from the cold Nordic Sea. The passage of time so evident here is also the perfect leitmotif for Celsius' life and scientific method. He took the long view in everything he did, able

Olof Rudbeck the Elder (1630–1702) – the eccentric 'universal genius' of Celsius' home city and university in Uppsala. This illustration from his 1689 book *Atlantica* depicts him revealing Scandinavia's underworld, surrounded by classical figures including Plato, Aristotle, Odysseus, Ptolemy, Plutarch and Orpheus.

to wrench off the blinkers of the present to see and understand the world around him in its true context of ages long past and yet to come.

Throughout his career, Celsius' sights were fixed upon how his work could practically benefit humanity in the years that followed. In his 1733 account of observing the Northern Lights across Europe, he wrote: 'It will bestow on our century a greater honour to have left true observations to later generations, rather than false hypotheses which can easily be refuted.'[2]

In geological terms, Uppland was underwater until quite recently. During the Neolithic period (4500–1900 BCE), the sea level was effectively 30 metres higher than today. And when the Vikings lived here (800–1050 CE), it was still some 7 metres higher. Waves once lapped the boulders surrounding the burial mounds now far inland, making it possible for heavy, clinker-built boats to be hauled ashore and entombed with the revered owners and their riches, beneath the man-made hills.

A Viking ship passing burial mounds on the River Fyris. Drawing by Olof Thunman.

Today, standing slightly apart from the church, an angular wooden bell tower juts dramatically towards the heavens, its shuttered openings and shingle-covered shape evoking a gigantic, armour-clad warrior. Coated in the region's distinctive, iron red *falu rödfärg* paint and the petrified pitch of many decades, the tower looks out over the curving chain of burial mounds: a glowering sentinel silently registering the passage of generations.

As I followed the twisting path between the mounds, I pondered Celsius' contribution to solving the riddle of this upsprung place. It was a profound, slowly unfolding truth that occupied half of his working life. The natural history that Celsius helped to reveal about his homeland was the perfect example of his experimental dexterity, daring imagination and urge to collaborate. His sharp eyes and singular mind seized upon the dynamic nature of the universe and our place within it.

From the top of the Högåsen ridge of burial mounds, I had an uninterrupted view across bright yellow fields of rape to Uppsala, a few kilometres away, its sixteenth-century castle and the tapering shards of the cathedral's towers standing out in stark silhouette. It was at the university there that Celsius became recognised as a thinker of rare originality, there that he built up the reputation that saw him welcomed in the great European centres of his age to study eclipses of the sun and the aurora borealis. And it was to Uppsala that he returned in 1737, after the Arctic expedition that brought him to prominence and financial security.

In his final years, Celsius established Sweden's first astronomical observatory in the centre of Uppsala, where he assembled a powerful coterie of skilled disciples and made further, revolutionary discoveries about the nature of light, temperature and the Earth's magnetic field. He was a man in obsessive pursuit of primal principles, a relentless empiricist for whom scientific enquiry and evidence were everything.

One can imagine such a driven and single-minded person being somewhat awkward or oppressive company. But Celsius' own writings, and the memories of those who knew and worked with him, suggest not. He was, by all accounts, eloquent, polite, sensitive and appreciative of others, calm, easy-going and well liked. Celsius could sing, play the guitar, and was an entertaining conversationalist. A source from the Swedish National Archives[3] describes him as 'a versatile man. He wrote Swedish and Latin verses. Always happy and cheerful, no matter how busy he was, he never seemed to be in a hurry.'

A generation later, another Uppsala astronomer, Bengt Ferner, met with the sister of Joseph Nicolas Delisle, a French colleague with whose family Celsius stayed while he studied at l'Académie des Sciences in Paris. Ferner wrote how: 'The [by then] old woman talked about him with ecstasy ... It is extraordinary how Celsius has been able to just fascinate people into loving him.'[4]

Monarchs, institutions and peers sought out the unassuming Swede, whose life coincided with those of some of the most renowned names of the era. His contemporaries included Carl Linnaeus, Leonard Euler and Benjamin Franklin: titanic figures of the Enlightenment who, like him, grappled with the biggest questions about the universe and natural world.

Celsius' career also overlapped with that of his fellow pioneer in the field of thermometry, Daniel Gabriel Fahrenheit (1686–1736). While every good school student knows that the two men's respective scales come into perfect alignment at a single point – minus 40 degrees – the scientists themselves never met. And neither of them lived long enough to see their inventions assume lasting worldwide importance and use.

Humans are one of only a handful of animals that have learned to make and use tools. And the most valuable tool we possess in the twenty-first century to respond to the existential emergency of climate change is the manipulation of data – collecting, storing and interpreting information about what is happening around us. Our ingenuity is being put to the test: how can we apply and act on this knowledge to find more sustainable ways to live in harmony with the natural world? It is exactly the kind of problem that captivated Celsius. If he were alive today, he would surely be in the phalanx of Earth sciences, wielding his paradigm-shifting theories and evidence to find the best responses.

Before Celsius pioneered his meticulous methods of logging and analysing records of weather, astronomical and geophysical events, such an approach would have appeared as alien to his peers as alchemy or necromancy seem to us today. We take it for granted now that, to comprehend something, find solutions and make projections, we explore, inspect, research, measure and collate findings. And, in the age of big data, it is self-evident that the more expansive and finer-grained information is, the more insight it can potentially yield. But these are quite recent practices – ones perfected and first applied to climate studies by the man whose name we all recognise, but about whom most know so little: Anders Celsius.

Some changes are either too fast, too massive or too gradual for us to perceive from our personal, *terra firma* standpoint of the now. Like the beating wings of an insect, a bullet fired from a gun, watching plants grow or the hour hand of a timepiece move around the clockface, we cannot see these things happening, but the proof that they are confronts us every time we glance back. Celsius understood that the tiniest movements recorded by the brass, wood, glass and alcohol of his custom-crafted instruments indicated vast effects taking place over aeons. The sorts of information he collected and curated became the rich and elegant language through which we can describe such things, and accurately predict what will happen next.

Celsius was a progressive and contented internationalist, equally at ease in his native country or among learned colleagues, aristocrats and royalty in Italy, France, Germany or Britain. He was a man far ahead of his time, who understood that natural changes wrought over thousands or millions of years pay no heed to the arbitrary boundaries and orthodoxies through which humans attempt to impose order. He also grasped a new and thrilling concept: that the most successful responses to these phenomena might be realised through the aggregated effects of a multitude of smaller actions.

Unlike many of his contemporaries and peers, Celsius is now little remembered and celebrated beyond the graduations etched onto a thermometer or gauge. The reasons why lie partly in Celsius' lineage and the rancorous disputes between nations, church and state during the golden age of discovery. The intricate, constantly shifting religious schisms and geo-politics of competing states in eighteenth-century Europe shaped Celsius' ambitions and opportunities. But his own enigmatic personality and the setting and culture in which he grew up also affect the way he's remembered now.

Some people are famed for one overriding idea, invention or event. But Celsius' contributions were more fragmented and broadly spread. His adventures and achievements are not just arresting in their own right, but also for what they reveal about the process of scientific insight and the trials of academic collaboration, criticism and competition. And as to Celsius' relevance now, we need only observe humanity's impact upon the Earth and witness its signs of pain. I contend that understanding the mind and methods of Anders Celsius can help us to ask the right questions about our current predicament, and perhaps find some answers.

2

FORGED IN FLAMES

The Origin Story of Celsius the Scientist (1701–02)

There is no education like adversity.

Attrib. Benjamin Disraeli

It was the smell they noticed first. Shortly after midnight on 16 May 1702, a tart whiff of woodsmoke and melting pitch entered the Celsius family house in the centre of Uppsala. Propelled by a stiff Baltic breeze, it seeped invisibly through the eaves and windows, permeating the interior with a pungent sense of danger. Head of the household, 44-year-old Nils, woke with a start, the acrid odour catching at the back of his throat. Around the edges of the painted wooden bedroom shutters, he saw a faint orange glimmer. His wife Gunilla stirred beside him, she too now roused by the bitter aroma that had invaded their home. A quick glance from the window confirmed what they feared: their city was on fire.[1]

Scooping up their 6-month-old son, Anders, and 6-year-old daughter, Sara Märta, and venturing outside, Nils and Gunilla immediately realised that this was no ordinary fire. The whole sky was lit up in a ferocious, dancing scarlet. Overhead, birds disorientated and driven from their roosts by the inferno, zig-zagged in confusion, their undersides flashing with reflected, iridescent shades of pink and orange.

And now the frightened family registered the sound: a deep roar, growing by the second into an angry howl as the flames consumed everything in their path. It was an elemental force unleashed, its voice growing in belligerence with each structure that fell. At every corner, courtyard and alley, the wooden buildings ignited, intensifying the terrible heat and power of the blaze.

There was no choice but to join the flood of other anxious households streaming out of the city. There were families like them, pushing carts piled high with hastily gathered possessions, as well as stooped figures of older residents, plus merchants and academics with their servants, and an occasional weeping child, distraught at being separated from its parents. Animals too: dogs, rats, cats, goats, hens – even droves of pigs being herded north, the fretful owners flicking sticks at their hairy rumps.

As Nils and Gunilla turned at the end of the street, clutching their children close against the flying ash and embers, they saw an animated line of their fellow citizens silhouetted against the glow. The desperate neighbours had formed a human chain to the riverbank, dropping leather buckets down into the River Fyris on ropes to be hauled up and passed, hand to hand, towards the fire then back to the river. It was frantic and brave, but futile.

Others took to small boats – hurling their belongings on board and attempting to flee the city by heading downstream towards Stockholm. But they ran the gauntlet of burning trees, falling bridges and building timbers crashing into the water around them. Wall by wall and roof by roof Uppsala was being destroyed and, with it, the prospects and prized possessions of most of its 5,000 or so inhabitants. The scorched fragments of all these lives swirled in the wind as the escapees fanned out into the countryside.

When the Celsius family was able to halt at a safe distance and gaze back at the conflagration, it must have looked like the end of days. In just a few hours, so much that was familiar and stable in their world had vanished. Having lost almost everything they owned, how could they start again and build a future in a city that lay in smoking ruins?

Baby Anders, born the previous November, did not and could not know it then, but for him this was a beginning. His parents' alertness had not

Woodcut of the 1702 fire that destroyed the city of Uppsala, by Olof Rudbeck the Elder.

only spared his and his sister's lives, but also marked out his academic path ahead, and they had managed to save most of the precious astronomy and mathematics books belonging to Gunilla's late father, Anders Spole. These volumes would become the bedrock of the young Celsius' learning.

No one is sure exactly what caused the 1702 Great Fire of Uppsala, but it is thought to have started near the main square and town hall, which also housed the city's board and court. Two neighbours nearby, Professor Upmarck and Academy Rent Master Rommel, had been locked in a long-running dispute over the boundary between their plots. And speculation soon arose that this might have been responsible. Perhaps the simmering enmity between these two prominent men boiled over into arson by one of them, or maybe it was an ill-considered, attention-seeking protest that got out of hand. Either way, the official investigation into the fire in 1704 was unable to determine a specific cause.

But while its origins were unclear, both then and now, its effects were obvious. Apart from sections of the sixteenth-century castle and the cathedral's stone core,[2] most of the city centre was obliterated. But the university's library and Gustavianum main building[3] survived largely intact. Accounts report that, at the height of the fire, a lone male figure was seen standing on its roof, silhouetted against the squat dome that topped the magnificent 200-seat anatomical theatre, one of the finest in Europe. The man's long hair streamed out behind him as he boomed out instructions to direct the frenzied efforts below. It was Olof Rudbeck – Uppsala's famed, 72-year-old sage of science, art and humanities.[4]

Rudbeck had designed and overseen construction of the dome behind him and created the city's botanical garden – second only to that in Paris – in which the rare and exotic plants had already been reduced to cinders. Despite also losing his home and most of his own substantial book collection in the fire, within a few days of the destruction Rudbeck presented himself to the Uppsala Council with a scroll of plans for the city's rebirth. The Council adopted many of his ideas, but he never got to see the vision fulfilled. Rudbeck died just four months later, his lungs seared from his valiant rooftop exploits.

In times when guttering tallow candles and lamps were the principal sources of artificial light, urban fires were commonplace. Back to first-century Rome, Sweden's capital Stockholm in 1625 and, most famously, London in 1666, it was not unusual for whole cities to be devastated.

Uppsala Cathedral with the Gustavianum University building in the foreground, c. 1900.

Uppsala had suffered a similar fate before in 1437 – and in parts, would do so twice more in 1743 and 1809. Before the existence of organised municipal services, firefighting resources and techniques were rudimentary at best. In larger places, the authorities sometimes ordered whole streets or districts to be pulled down to create breaks to check the spread of fire. But in Uppsala in 1702, things were both too compact and too quick for this tactic. Once the first few rows of houses were ablaze, and with the wind driving the flames, the fragile wooden city was doomed.

But as its landmarks and treasures disappeared in showers of sparks, good fortune also settled over Uppsala that night. There were apparently no human fatalities at all – each household replicating the lucky escape of the Celsius family. People were able to detect the fire and raise the alarm just in time to dodge death and flee the danger area before they were overcome. But there were many casualties of a different kind: farm, draft and domestic animals left behind, plus wild rodents, reptiles, spiders and insects. Countless individuals from other species perished unseen and unmourned. The fire was a foretaste of far greater global calamity to come.

Anyone visiting the leafy university city today will find few obvious signs of the devastation. Over the decades following the 1702 fire, corresponding roughly to Anders Celsius' lifespan and presence there, Uppsala was reconstructed on a new and more regular grid street pattern, which drew heavily on Rudbeck's ideas. The cathedral was restored, albeit without its medieval interior decorations and flying buttresses, while the castle was remodelled onto a smaller footprint, minus a whole wing. But a proud scholarly soul and the determined nature of Uppsala's inhabitants remained intact. The city and its university rose again.

At the top of the steep escarpment next to the castle a large bell now hangs, suspended within an open framework of hefty timbers bound with solid iron fastenings. Appropriately, and by coincidence, it is named the Gunillaklockan (Gunilla bell), after Sweden's sixteenth-century queen rather than Celsius' mother. But it is a fitting (if unintentional) tribute to one of the fire's innumerable heroines who, with her husband, safeguarded their young son so that his genius could later be realised. The Gunillaklockan is poised to ring out if Uppsala faces another crisis.

Of course there *is* now another emergency threatening not just Uppsala but the Earth itself. It is a slow-burn catastrophe of unimaginable dimensions, signalled each summer by worsening mega-fires in the forests of

A pre-Great Fire view of Uppsala Castle, Cathedral and the Gustavianum dome, looking east towards the city centre.

Uppsala's Gunillaklockan tower overlooking the city from the castle.

every continent. By contrast to the events of 1702, the reasons for these *are* clearly known and understood. And, unlike the peril that confronted the Celsius family and their neighbours, this time there is nowhere to escape to.

These are fires humanity has to face. And, if we are wise, we will draw inspiration from the little boy whose future was forged as he escaped the flames in his parents' arms that May night.

3

A CITY OF LEARNING

Uppsala and Sweden's Age of Freedom (1720–72)

A man travels the world over in search of what he needs and returns home to find it.

George A. Moore, *The Brooke Kerith: A Syrian Story*, 1916

Aside from his family's escape from the Great Fire of Uppsala, a curious combination of other people, places and events influenced Anders Celsius' birth and early life. It was a three-generation fusion of ancestry, state, royalty, religion, geography, academia, conflict and coincidence that set him on his path to greatness. The Sweden he was born into on 27 November 1701 was a country unbalanced by a struggle for supremacy between Church and Crown and weakened by successive wars with neighbouring Denmark and its allies. Yet the nation was about to emerge into its golden Age of Freedom – an era when bright minds and busy spirits like Celsius could rise up and shine in the European Enlightenment.

It was the complex sovereignty, religion and politics of the time that created this lineage and transition, which placed Uppsala at the very centre of national affairs. How this all came to be illustrates the way knowledge travels across time, carried by humans and passed on through what they leave behind. Celsius' formative years were like one of the intricate patterns carved into the ancient rune stones that still grace the Scandinavian and Nordic landscape. Just as the mysteries and whispers of genealogy in these obelisks stretch forward into the future, Celsius' familial and intellectual heritage was firmly etched into his fortunes, to reveal their benefits later on. Gaining insight into today's great questions from Celsius' life presents a similar challenge, to understand and interpret what we see before us, and decipher what it means for times yet to come.

Uppsala University was founded in 1477. Two years earlier, the Swedish government learned of a visit to Rome by Denmark's haughty and go-getting Queen Dorothea of Brandenburg. There, she had persuaded Pope Sixtus IV, a strong patron of the arts and learning, to issue a papal bull, giving permission to create a university in the Danish capital.

Desperate to avoid being outdone by the Danes, and keen to claim the title of the Nordic region's first fully fledged seat of higher learning, Jakob Ulvsson, Archbishop of Uppsala and Primate of the Catholic Church in Sweden, acted quickly. Apart from his own close ties there, Uppsala was the archbishop's obvious choice because it had been one of the country's most important episcopal sees since Christianity first reached this part of Scandinavia in the ninth century. It was the perfect place for Sweden to beat its Baltic neighbour to pole position in the late-medieval brains race.

Uppsala had grown up from a cluster of old settlements around the River Fyris, which flowed down to the port capital of Stockholm. The city was a busy, thriving hub for business and agricultural trade – the university just *had* to be there, Ulvsson reasoned. So he called on his ecclesiastical contacts and others who owed him favours to secure his own papal

Key figures in founding Uppsala University in 1477: Archbishop Jakob Ulvsson (*c.* 1430–1521) and Pope Sixtus IV (1414–84). Statue at Uppsala Cathedral and portrait by Pedro Berruguete.

decree from Sixtus IV, to bless the project and confer its corporate rights.

The Vatican's endorsement came with certain benefits and conditions attached. Uppsala would immediately gain status equal to that of the University of Bologna in northern Italy – the world's first and longest continuously operating, degree-awarding institute, founded in 1088. The whole idea of a university had come from Bologna; growing out of mutual aid societies (*universitates scholarium*) in which foreign students hired local scholars to teach them and grouped together to protect themselves from sanctions arising from any misdeeds or debts of their predecessors.

Being put on a par with this venerable centre gave Uppsala the right to create four faculties (theology, law, philosophy and medicine), and to award bachelor's, master's and doctoral degrees. From the outset, there was also teaching and research in astronomy, although the subject then was not as we understand it today. True to its classical roots, study of the cosmos was still bound up with the moral and spiritual issues of natural philosophy and religion – seeking better to understand the stars as a way to live and worship God as Creator.

By way of personal recognition, Archbishop Ulvsson was appointed the university's first chancellor. And best of all – from the Swedish point of view – this was all achieved two years before Copenhagen University was ready to open its own doors in 1479. In the long-running rivalry between the neighbouring states, Sweden had nudged ahead.[1]

Uppsala University had a modest beginning, with just a few dozen students and a handful of scholars. But its confidence and activities steadily increased, turning the city into a lively place of intellectual excellence to match its religious and commercial credentials. The University continued to expand and enhance its standing until, in the sixteenth and seventeenth centuries, it ran into the turbulent wash of Sweden's break with the Catholic Church. This split would have serious consequences for the ancestors of Anders Celsius.

The Swedish Reformation to become a Protestant country based on the new and progressive tenets of Lutheranism was a stuttering, protracted affair lasting over sixty years.[2] It began in 1515 and saw bloody massacres, the suppression of monasteries, excommunications, the abdication of a queen and a union with Denmark. During this period, Uppsala University fell into torpor and decline. But while its early energy and promise were halted by swirling disputes between competing royal, religious and

King Charles IX of Sweden (1550–1611) – champion of the Protestant cause and State.

political blocs, the city itself remained important. And through the Synod of Uppsala in 1593, it played a decisive role in the Swedish Church breaking its final ties with Rome.

On 1 March, Duke Charles (heir to the Swedish throne and later King Charles IX) summoned four bishops and 300 priests in all their regalia to Uppsala for an official confession of faith. The duke entreated this gathering to adopt the Lutheran Augsburg Confession – unaltered and in its entirety.[3] Three weeks later, as the unanimous vote passed and the decrees were signed, the Synod's elated chair, theology professor Nicolaus Olai Bothniensis, declared: 'Now Sweden is one man, and we all have one Lord and God.'

The country's shift to Protestantism was a turning point for the University. It was swiftly rejuvenated to serve both the State and the Church of Sweden – in particular, to educate future generations of ministers. To mark this rebirth, a full-time professorship of astronomy was also founded – the first of a raft of new roles, statutes and buildings (including a dedicated university library) that followed in the early part of the seventeenth century.

In 1624, King Gustaf II Adolf continued on the reforming path. He decided that ownership of 300 Crown Estate farms spread across eastern Sweden should be passed to Uppsala University. These were well-established and productive assets, so at a stroke, the King's gift secured not just financial stability for the institution, but also a measure of independence. Notwithstanding yearly fluctuations in harvests and volatile produce prices, by the mid-1600s the wealth generated by the farms saw the University's student numbers swell to more than 1,000 – roughly a third of the city's whole population. And a century later, Anders Celsius' family would directly benefit from the fecundity and produce of these tied smallholdings.

4

A FAMILY OF AMBITION

The Celsius Name and Prominent Ancestors

People will not look forward to posterity who never look backward to their ancestors.

Edmund Burke, *Reflections on the Revolution in France*, 1790

The first Swedish scientist to bear the name Celsius was Anders' paternal grandfather, Magnus. He was born on 16 January 1621, the son of Nicolaus Magni Travillagaeus, a farmer-turned-church minister from Östergötland in south-eastern Sweden. Magnus grew up in Alfta, a village in the Ovanåker municipality about 150 kilometres north of Uppsala. As Magnus prepared to leave home and begin his university studies in 1641, he decided to follow the vogue for new scholars adopting Latinised names. First, he chose Helsingius (meaning 'unique' or 'visionary' and linked to the name of his home province), then switched to Metagrius ('beyond wilderness'), before eventually settling on Celsius.[1]

Inspiration for this final choice came from the vicarage in which he lived, Högen (meaning 'hill' or 'highpoint'). In a bold declaration of self-belief, Magnus attached himself to exalted and higher powers – as in the Christian praise *'Gloria, hosanna in excelsis'* – 'Glory to God, salvation in the highest'. It was a clear and public statement of erudition, upon which he was determined to deliver.

Magnus Celsius arrived in Uppsala at a time when student life was reaching new and daring heights of exploration. He found a heady mix of academic and social freedoms that often – with the addition of a third element, alcohol – spilled over into rowdy incidents and riotous conduct. Unsurprisingly, the Church authorities bristled with disapproval at such scenes. As both a deterrent and punishment, they created the *prubba* or *proban* in the courtyard of the main university building – a rough, lock-up detention house, where the worst offenders were incarcerated to sober up, reflect on their wrongdoings and await further censure or sanction.

A century later, the Swedish poet Anna Maria Lenngren (1754–1817) satirised students' unruly manners in her class-conscious poem 'Biographie'.[2] She imagined a middle-aged man reminiscing about the wayward excesses of his scholarly youth:

> But learning does its labours have,
> So as a student, I was quite persistent,
> About the city folk, small pamphlets I wrote
> Went roaming wild on the streets each night,
> And with the fanatical journeymen did fight,
> In the Proban you could often see me –
> And with the greatest honour I took my degree.

There is no record of Magnus Celsius indulging in such behaviour or enduring this sort of penalty. His correspondence and academic records from the time suggest he was something of a model student: gifted in mathematics and astronomy, and showing an early interest in the subject that was to occupy much of his later years – deciphering the ancient, staveless alphabet of Sweden's rune stones.[3] He was an accomplished mechanic, craftsman and artist too, building his own astrolabes and telescopes, as well as painting, printmaking, engraving and creating sculptures and poetry.

Magnus graduated with a doctor's degree in 1649 and became headmaster of Uppsala School in 1656. Nine years later, he was made temporary professor of astronomy, and three years after that he was elevated to a full professorship in mathematics. The first-generation Celsius had made his mark.

To further his research and fascination with runes, Magnus also took up a role with the National Board of Antiquities – a farsighted venture with the enthusiastic personal support of King Gustaf II Adolf (1594–1632), to protect and preserve the country's most precious monuments, artefacts and chronicles. Founded in 1630, it was one of Europe's first initiatives to safeguard and document historical treasures, which created the blueprint for the heritage bodies that exist in most countries today.

A new endeavour like this demanded vision, determination and strong powers of persuasion. These are all qualities detectable in the likeness of Magnus Celsius captured by an unknown artist in 1661. The 40-year-old professor sits erect and confident, his chiselled features framed by a perfectly starched, square, white collar and lush brown hair cascading over his shoulders. He has a strong brow, firm jaw and prominent straight nose, with a pencil-thin, drooping moustache and a neat tuft of beard below his lower lip. It is the poised countenance of a man at the peak of his powers – someone assured in his views and ability to take on whatever or whoever stood in his way.

A few years before he sat for this portrait, Magnus had married Sara Figrelius, the 19-year-old sister of a fellow professor at Uppsala University, Edmund Figrelius, who specialised in history and philology (the structure and development of languages). The couple had five children: three sons (Nils born in 1658, then Johan in 1660 and Olof in 1670) and two daughters (Katarina in 1666 and Barbro in 1670). Little information survives about the two girls, but all three boys achieved distinction in their chosen professions.

Nils followed his father into astronomy, and so paved the way for his own son Anders to do the same. Johan showed considerable early promise as a dramatic playwright, authoring and staging classically inspired pieces with the student troupe Dän Swänska Theatren at the Lion's Den in

The only known likeness of Celsius' paternal grandmother, Sara Figrelius (1636–1720).

Stockholm, one of the capital's first public performance spaces. And in later life he became a jurist and civil servant.

Meanwhile, Olof had a passion for botany, shared his Magnus' fascination with Sweden's runic languages and became professor of history at Uppsala University. In 1748 he published *Hierobotanicon*, a comprehensive account of all the plants mentioned in the Bible.[4] Olof would also play important part in the life and success of another famous Uppsala scientist, Carl Linnaeus.

In between parental duties, teaching, historical research and other pastimes, Magnus turned his hand to architecture. Inspired by the example of one of his own teachers, Bengt Hedraeus, he designed and built a small observatory on the roof of the family's house. From there, he used his astronomical observations to produce texts about chronology and calendars, as well as annual almanacs for sale to the public.

In 1677 yet another dimension opened up in Magnus Celsius' life, when he was made vicar of the parish church at Gamla Uppsala – the current cathedral's precursor that lies a few kilometres north among the ancient royal burial mounds. At the time of Magnus' arrival, his university peer, Olof Rudbeck, was in the early stages of an obsession to prove that, as well being the seat of Swedish religion and culture, this place had once been the epicentre of the lost continent of Atlantis.[5]

The spot where the modest church now stood, Rudbeck speculated, was formerly the island's capital. And so immense were his powers of logic and deduction that he was able to meld any evidence to fit this whimsical theory and counter anything that might contradict it. Rudbeck devoted the last twenty years of his life to chasing this myth.

Whatever Magnus Celsius thought of his colleague's beliefs, his time and tenure in Gamla Uppsala were short. He died of acute indigestion in 1679, survived by his children and his wife Sara, who lived for another forty-one years. But as one astronomer departed, another emerged. Magnus' eldest son Nils was set on following in his father's footsteps. And, in so doing, he would prepare the ground for his son Anders' later endeavours.

Anders Celsius had equally illustrious heredity on his mother's side. His maternal grandfather was Anders Spole, the son of a blacksmith, born on a farm in Målen in central southern Sweden in June 1630. After studying at secondary school in nearby Jönköping, at the age of 22 he went to Greifswald on Germany's north-eastern Baltic coast to study mathematics.

This city's university was founded two decades before that in Uppsala and it became the first of Spole's academic destinations, which also saw him learn the military arts of fortification, munitions and navigation in Prussia and Saxony. He returned to Sweden in 1655 to continue studying mathematics at Uppsala University, while also preaching at a local church and privately educating the two sons of an army general, Baron Carl Sjöblad.

Spole's worldly manner and blend of academic and military knowledge evidently made a good impression, because in 1664 his employer asked him to chaperone his sons (then both still teenagers) on an educational tour of Europe. For the older boy, Erik, this three-year peregrination led to a commission and distinguished career in the British Royal Navy. While for Spole, it meant meeting and spending time with some of the continent's most renowned scientists. He got the opportunity to work alongside the Dutch mathematician and astronomer Christiaan Huygens (1629–95), and in England he met the University of Oxford scientists Robert Hooke (1635–1703), who had major improvements to the design and use of microscopes, and the eminent chemist/physicist Robert Boyle (1627–91). Then in Italy, he encountered the astronomers Giovanni Riccioli (1598–1671) and Giovanni Cassini (1625–1712), the latter of whom went on to establish the magnificent Royal Observatory in Paris.

Learning new tongues and avidly acquiring books as he went, this tour was the making of Anders Spole. It was an intensive education by immersion that burnished his own character and intellect, as well as that of his young companions. On arrival in England, he attempted to master the language in a week by lodging with a local family – no doubt a rigorous trial for both the hosts and student. Spole also walked all the way from the French coast to Paris in ten days and took a voyage to Sicily to view the smouldering volcanic cone of Mount Etna. Like his contemporary and colleague Magnus Celsius, Spole proved himself to be ambitious, single-minded and energetic – traits that would surface afresh two generations later.

On his return to Sweden in 1667, Spole's exploits earned him a position as the first professor of astronomy at the newly founded Lund University, near to the southern tip of Sweden. Within a couple of years, he married Marta Lindelia – a bride who, like Magnus Celsius' wife Sara, was some twenty years his junior. They had a daughter, Gunilla, in 1672 and went on to produce eight more children over the next two decades.

As his family steadily expanded, the capable and diplomatic Spole's career blossomed. He became principal of Lund University in 1672 and continued in that role for four years until the institution was forced to close temporarily, because of a flare-up in the Northern Wars between the states that now make up Norway, Denmark, Germany, Poland, Russia and Sweden. This conflict gave the professor an opportunity to put his martial know-how into practice. He was enlisted to help defend the fortress at Jönköping, and then fought alongside King Charles XI at the Battle of Landskrona in July 1677.

In this encounter, a 13,000-strong Swedish force of regular army, peasant infantry, cavalry and artillery faced a similar-sized Danish army commanded by 24-year-old Christian V, whose kingdom was clearly visible across the choppy grey waters of the Øresund Strait. After some initial exchanges, a confused pitched battle broke out, leaving almost 4,500 troops from both sides killed or captured.

It was a clear although inconsequential victory for Sweden, and another turning point for Spole. Now aged 47, this clash on the tussocky Ylleshed Moor allowed him to add proven courage and tenacity to his undoubted intelligence and conscientious temperament. Two years after he emerged unscathed from combat, the lives and destinies of the Spole and Celsius families were about to come together.

5

A FRUSTRATED FATHER

Nils Celsius and his Ruinous Dispute
with the Church of Sweden

> Someone once said that every man is trying to live up to his father's expectations or make up for their father's mistakes.
>
> Barack Obama, *The Audacity of Hope: Thoughts on Reclaiming the American Dream*, 2006

In 1679 Anders Spole moved his family north so he could take up the professorship in astronomy at Uppsala University. Shortly after he began his new duties, a 21-year-old student approached him, seeking help to submit his dissertation for oral defence. It was Nils Celsius, the eldest son of Magnus, who, with guidance from his recently deceased father, had written a thesis '*De principiis astronomicis propriis*' ('On the Principles of Proper Astronomy').[1] With his father now dead, Nils needed another senior academic to support him. And in the newly appointed Spole he spotted the ideal, high-profile candidate.

An engraving by Elias Brenner from around this time shows what the hopeful Nils must have perceived in the studious professor. A stern but learned man stares pointedly at the viewer, his half-frown suggesting both superiority and intent. A luscious dark wig tumbles around his shoulders and he wears a deeply folded satin tunic. Here was a man of serious status and repute.

Spole recognised the supplicant student's name and ancestry of course, so looked kindly on the recently bereaved young man. And, little realising the explosive potential of what lay within the dissertation's nineteen pages, he agreed to help.

Anders Spole at the peak of his academic career as Uppsala's professor of astronomy. Engraving by Elias Brenner (1647–1717).

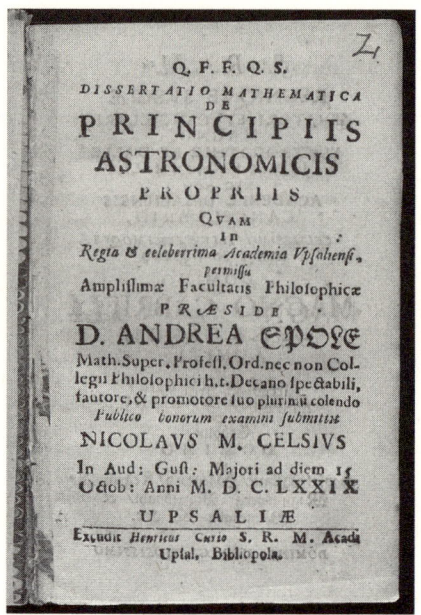

Highly volatile material – the front page of Nils Celsius' 1679 astronomy dissertation, which drew him and his future father-in-law, Anders Spole, into a bitter dispute with the Church of Sweden.

Relationships between the University and the Church of Sweden had been strained and simmering for years – not just because of the behaviour of students who found themselves in the *prubba*, but also owing to the increasing influence of non-conformist thought that challenged literal interpretations of the Bible. Theologians found their religious absolutism assaulted on all sides – by Nicolaus Copernicus' sixteenth-century model of the known universe (with the Sun at its centre rather than the Earth)[2] and by René Descartes' more recent philosophical treatises about the nature of truth and how to obtain reliable knowledge.[3] Descartes had died in Sweden in 1650, while briefly working as a private tutor to Queen Christina. For the country's senior churchmen, Cartesian thought and other revolutionary ideas emerging from the very universities they had founded were getting alarmingly close and discomfiting.

Nils Celsius' 1679 dissertation struck these already troubled waters like a meteorite. Its impact threatened to end his academic career before it had begun, and the radiating shockwaves of discord continued for the rest of his often unhappy and unfulfilled life. With Spole's backing, the exposition was printed and published, provoking an immediate storm of outrage. In it, Nils declared that astronomy should be 'based upon theories, hypotheses and observations, leading to conclusions free from prejudice or preconception'. And he clearly had clerics in mind when he wrote that treating the contents of the Bible as indisputable fact 'puts a mask of pretended piety over the truth'.

This was too much for the University's ecclesiastical leadership. They insisted that the oral defence be postponed and summoned both the student and his supervising professor to a face-to-face interrogation. By now alert to the serious trouble brewing, Spole politely explained that he could not attend the hearing because he had broken his foot and could not walk. Convenient, true, both or otherwise, this tactic bought the pair some time, but the authorities were not to be deterred. They insisted instead that Nils Celsius provide them with a written justification of his claims.

Called to account for his audacious opinions, the student complied. But his response made things even worse. The theologians reacted with fury, demanding that Celsius must either remove the two most (to them) offensive pages of his dissertation or accept that his oral defence and degree be indefinitely cancelled. In response, Nils channelled the obstinate, uncompromising essence of his late father and refused to change a single word.

The row rumbled on for ten years, gradually spreading to infect the entire affiliation between Sweden's Church and all of its universities. It took the direct intervention of the King to bring peace and a new understanding between the factions. In 1689 Charles XI established a learned committee from both sides to reach a new settlement. Its members concocted a simple but elegant compromise: from now on, they decreed, scholars could enjoy complete intellectual liberty, but must at the same time pay due respect to the Church and its teachings. The proclamation's boundaries were fuzzy enough to allow both sides to function and save face. It presaged Sweden's Age of Freedom,[4] and enabled both Anders Spole and Nils Celsius to resume their work without the threat of religious opposition hanging over them.

What Spole did by now have above him was a newly constructed home observatory, complete with hand-crafted telescopes and instruments to scan the night sky and augment his teaching. From here, he watched Kirch's great comet of 1680,[5] its long, dusty silver tail slashing a dramatic arc almost perpendicular to the horizon.

The Great Comet of 1680 – the first to be discovered by telescope, and so bright as to be visible in daylight.

Spole also continued to invest in the future of the ill-starred Nils Celsius, who was still struggling to find a permanent job and was supporting his part-time lecturing at Uppsala in mathematics, astronomy and Latin by earning extra money as a land surveyor around the Baltic provinces. In a forthright gesture of support, Spole successfully recommended Nils for a royal scholarship at the University. And in 1691 the bond between the two men became even closer, when Nils (by then 33) married Spole's eldest daughter, Gunilla – another union of a 19-year-old woman with a much older academic.

A few years later, the King granted Spole and his colleague Johannes Bilberg (a cleric and former professor of mathematics at Uppsala University, who had also become embroiled in the row about Cartesian principles) an exciting and radical commission. They were directed to travel to Sweden's far north to investigate the phenomenon of the midnight sun, which brought continuous daylight to these extreme latitudes each summer. As an absolute monarch, Charles XI wanted to know and comprehend more about this annual spectacle and how it related to his own 'Sun King' status.

Scientific expeditions of this sort were then almost unknown in the Nordic world, so the two men eagerly accepted the task – artfully expanding their brief to also encompass studies of their own interests in light refraction, gravity, magnetism and the region's botany. For this last element they took with them the university's professor of medicine, 35-year-old Olof Rudbeck the Younger, son of the Atlantis-obsessed eccentric of the same name. After studying in the Netherlands, Rudbeck the Younger had returned to Uppsala to succeed his father three years before.

The group set off from Uppsala in late May 1695, tracing the Baltic coastline northward to the town of Tornio at the top of the Gulf of Bothnia. And from there they followed the Torne River valley 120 kilometres up to the village of Kengis. Even at this time of year, as they ventured further north, snow and ice hampered their progress. Bilberg described the hazards they faced in his journal:

> In the highway the ground being sometimes slippery, and the Horses and Charriots sticking in dirty Places, created us great Trouble, where the Cold being partly dissipated, yet the Earth did not return to its former Solidity and Firmness; for at this Season the Creeks nearest the Shore being harden'd with Frost and Ice, and many Pines being cut down, and disposed into order did even then represent a Winter Journey, the Frost being melted and thawing near the Bank-side only, so that no body cou'd go any further of the Journey with Safety.[6]

The expedition's astronomical findings were undermined by its short duration and inaccurate instruments, but Rudbeck's beautiful colour illustrations of the flora, fauna and landscapes they discovered made a much greater impression. Published in 1701 as *Lapponia illustrata*, his album of drawings became a landmark nature reference, and the work for which he is most remembered.[7]

Forty years later, Spole's grandson Anders Celsius would face similar obstacles and achieve fame by making his own expedition to the same place, on an even more ambitious quest: to determine the exact shape of the Earth.

King Charles XI of Sweden (1655–97) – when crowned at the age of 17, described as 'virtually afraid of everything, uneasy to talk to foreigners, and not daring to look anyone in the face,' but he became one of the country's most successful military commanders and revered monarchs.

Olof Rudbeck the Younger – Anders Spole's companion on his 1695 expedition to Lapland.

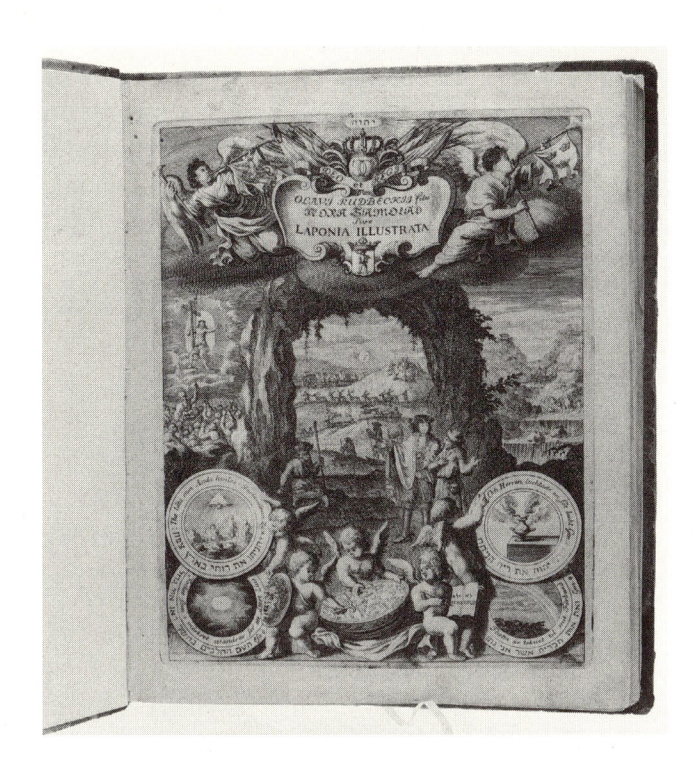

Frontispiece of Olof Rudbeck the Younger's *Lapponia Illustrata* (1701).

As the seventeenth century drew to a close, Spole's hearing deteriorated almost to the point of complete deafness and he reluctantly had to acknowledge that his professional life was over. In one last effort to secure his son-in-law's academic future, he wrote directly to the King to suggest that Nils Celsius should succeed him as professor of astronomy. And to finally bury the hatchet with the Church, he even persuaded the Archbishop of Uppsala to support the plan.

But old rivalries ran deep. When the proposal came before the University Senate in 1699, Nils Celsius' appointment was quashed by the influence of Olof Rudbeck the Elder – by now an aged veteran and expert manipulator of academic quarrels. The professorship went instead – by one vote – to the well-connected, 38-year-old staunch Cartesian, Petrus Elvius. The new professor quickly proved his worth – publishing ideas for a mechanical planetarium (*machina astronomica*) and helping to found the Collegium Curiosum, the embryo of the Swedish Royal Society of Sciences.

Spole died later that year, leaving behind his personal library of more than 1,000 volumes. These were the books saved from the Great Fire by Nils and Gunilla Celsius a few years later, which were to prove so important to his as-yet-unborn grandson Anders. After his death, Spole's family connections took another turn when Petrus Elvius married another of his daughters, Anna Maria. So, on top of their scholastic competition, Elvius and Nils Celsius became related by marriage. And when Elvius died in 1718 he was buried in his father-in-law Spole's grave in Uppsala Cathedral. The Latin inscription erected above them by his widow Anna Maria translates as:

> In eternal memory of objective, outspoken,
> talented and educated worthy father and husband,
> teachers of astronomy at Uppsala University.
> Beauty, glory, gold and wealth perish like lilies,
> but the soul, radiant in the light of God, will be forever.

By the time of his father Magnus' death, Nils Celsius was himself a parent. In December 1695, he and Gunilla had a daughter, who they called Sara-Märta in honour of their respective mothers. In November 1701, a son, Anders, followed, who they also ancestrally named in memory of Spole.

If Nils thought that the milestone of parenthood might mark the end of his youthful tribulations, he was wrong. Another son, Magnus, died as an infant, and within a year came the trauma and loss of the Great Fire. Then, three years later, his reputation took another blow when he made a critical mistake in compiling an almanac. In the period before he finally secured a permanent university post and salary, Nils had fallen back on this relatively undemanding but reliable source of income. But he miscalculated the timing of Easter by a week – a careless error that brought yet more discredit upon him.

Nils Celsius died in 1724, frustrated and often ostracised by his peers, having never recovered from the turmoil and infamy sparked by his student dissertation. A 1718 portrait by Jan Klopper shows Nils as a weary and haggard man. His face and clothes sag, as if burdened by all his troubles, and his right hand rests unconvincingly upon a page, holding a pair of ornate metal dividers. He looks defeated and disinterested. By the time he was finally made a professor the following year, Nils was already 60 years old, tired and with his capabilities in decline. But behind him in the painting is a glimpse of his legacy: a heavily framed celestial globe – the stage onto which his gifted son Anders would soon step with confidence and success.

6

THE AGE OF ENLIGHTENMENT

Young Celsius and Science in
Eighteenth-Century Europe (1710–19)

> Dare to know! Have courage to use your own reason!
> Immanuel Kant, *What Is Enlightenment?*, 1784

Celsius' life and short but brilliant career coincided with a pivotal period in the development of European thought, science and culture. Like the physical and astronomical phenomena he went on to study in adulthood, the determining experiences of his youth had their origins far in the past, and continued to resonate long after his premature death.

The period that straddles the 1700s is with good reason often called 'the long century' because of its far-reaching precursors and enduring influence that overrode the neat boundaries of the calendar. Opinions differ about the starting point for what became the Age of Reason. William Harvey's 1628 discovery of how blood circulated in the human body, the founding of the Royal Society in 1660, or the publication of Isaac Newton's *Principia Mathematica*[1] in 1686 can all lay claim to stirring the intellectual and philosophical ferment that led up to the French Revolution in 1789 and subsequent rise and fall of Napoleon Bonaparte. Anders Celsius was born into the midst of it all, as the Enlightenment was gathering momentum to change the face of the Western world forever. His ancestry and education reflected this earthquake of ideas, from which the aftershocks still continue today.

The roots of the Enlightenment lay in Renaissance humanism, which flourished across the fourteenth, fifteenth and sixteenth centuries. The Italian, Greek, French and German philosophers of this era reappraised classical principles – revisiting the greats of antiquity to develop the tenets of modern science, but with an accompanying religious purpose. Few of the key figures at this time were atheists; in fact, many of them were in holy orders – their mission was rather to purify and renew Christianity by exposing it to fresh rigour and examination. This cohort wanted to dispense with the conventions of medieval theology and go *ad fontes* (back to the sources) of the Old Testament in search of basic truths and codes for living.

The scientific revolution that followed emphasised observation, theory and evidence. No longer were thinkers content to find answers in faith, legend or tradition – they strove instead for proofs derived from the method first advocated in Europe by Roger Bacon in the thirteenth century. The empiricist approach was based on rational scepticism, hypothesis, induction, deduction and refinement. And from this, the core themes of the Enlightenment emerged: reason, individualism, fraternity, equality, progress, happiness, dignity and liberty – the keys to all of which were seen to reside in nature itself. In his 1689 *Essay Concerning Human Understanding*, the English philosopher John Locke wrote:

> Let us then suppose the mind to be, as we say, white paper, void of all characters, without any ideas: How comes it to be furnished? Whence comes it by that vast store which the busy and boundless fancy of man has painted on it with an almost endless variety? Whence has it all the MATERIALS of reason and knowledge? To this I answer, in one word, from EXPERIENCE. In that all our knowledge is founded; and from that it ultimately derives itself.[2]

Human nature was, in Locke's view, infinitely mutable – so knowledge should be gained through accumulated experience, not by seeking to access some outside truth or higher power.

Today, from a developed world perspective at least, it is hard to imagine the intoxicating sense of possibility created by this new framework of thought, nor the fear and resentment it provoked among traditionalists. For Celsius and his progressive contemporaries, the Enlightenment fostered a fresh and stimulating epoch of freedom, independence and limitless potential. But it shook others to the core of their beliefs, and instilled panic about what might come next and where this could lead nations and societies.

These competing currents were already in full flow by the time of Anders Celsius' birth in 1701. His father Nils had become embroiled in disharmony following the perceived frontal challenge to the Church's authority from his student dissertation. The family's loss of their home and most of their belongings in the 1702 Great Fire then threatened to push the household into permanent destitution. But displaying the resilience of human spirit that the Enlightenment also celebrated, they picked themselves up and moved on.

It was Celsius' mother Gunilla who seized the moment. She scraped together the means to open an informal dining house for students and scholars in Svartbäcksgatan, near to their former home. And little by little, as the burnt city was slowly rebuilt around them, the restaurant grew in popularity.

Gunilla used meat, poultry, vegetables and dairy produce from the University's former Crown Estate farms to serve up inexpensive, wholesome fare in an easy-going atmosphere where academics could share their latest deliberations and experiments. Professors, scholars and students sat side by side on the eatery's benches, their differences in status temporarily suspended or ignored as they consumed the reliable daily menu that Gunilla conjured from the wood-fired oven-cum-stove. The mood was warm and welcoming – full tables, a lively hubbub of conversation and the clatter of cutlery on blue and white tin-glazed earthenware plates and bowls the signs of enterprising success.

There are no written records of exactly what Gunilla cooked, but the 1993 archaeological excavation of an eighteenth-century inn on the island of Koffsan in nearby Lake Mälaren gives some clues to what might have been on offer. There, travellers enjoyed hearty broths, stews, soups and sausages made with veal, rabbit, pork, duck, chicken or local fish. Characteristic regional flavours came from tangy capers, foraged herbs, nutmeg, cloves and juniper, with onions, carrots and beetroot slathered in buttery sauces. And for those with a sweet tooth there was apple tart, pears in syrup and carrot cake – all washed down with strong dark beer, fiery schnapps and tea brewed from freshly gathered dandelions or stinging nettles. This was Scandinavian soul food that also nourished the mind.

It is likely that Gunilla served very similar dishes – her repertoire steadily expanding as the establishment's reputation grew. Once her daughter Sara-Märta was old enough, she too became part of the enterprise: serving, cleaning and helping to feed this ready local market and secure the restaurant's niche in the city and university scene. And for as long as Nils struggled to command a regular salary, the eatery provided most of the

The preserved eighteenth-century kitchen in a house near to Gunilla's restaurant in the centre of Uppsala – she and her daughter Sara-Märta would have worked with amenities like these.

family's income. Crucially, the business yielded enough funds to pay for Anders' education.

To begin with, the young Celsius received basic instruction from his father, his uncle Petrus Elvius (Nils' victorious rival to the professorship of astronomy) and Nils' eventual successor in that post, Eric Burman. This potent triumvirate clearly awakened a thirst for knowledge in the boy's mind, greatly helped by the remaining titles from his grandfather Spole's academic collection. While many books had survived the fire, Nils and Gunilla had been forced to sell some of the collection the following year. But they held onto a number of crucial volumes – enough to captivate their clever young son.

In a lavish commemorative address for Anders Celsius in 1743, Swedish Minister of State Baron Anders Johan Höpken recalled how, from a young age, the boy had amused himself with the geometric figures he found within grandfather Spole's books – endlessly sketching circles, polygons and elaborate polyhedra to explore their intricate forms and hidden properties. Höpken claimed: 'He was not yet twelve years old when he first saw and immediately understood Bilberg's geometry.'[3]

The Bilberg referred to in the eulogy was Johannes Bilberg – Anders Spole's companion on the expedition to Lapland in 1695. In the

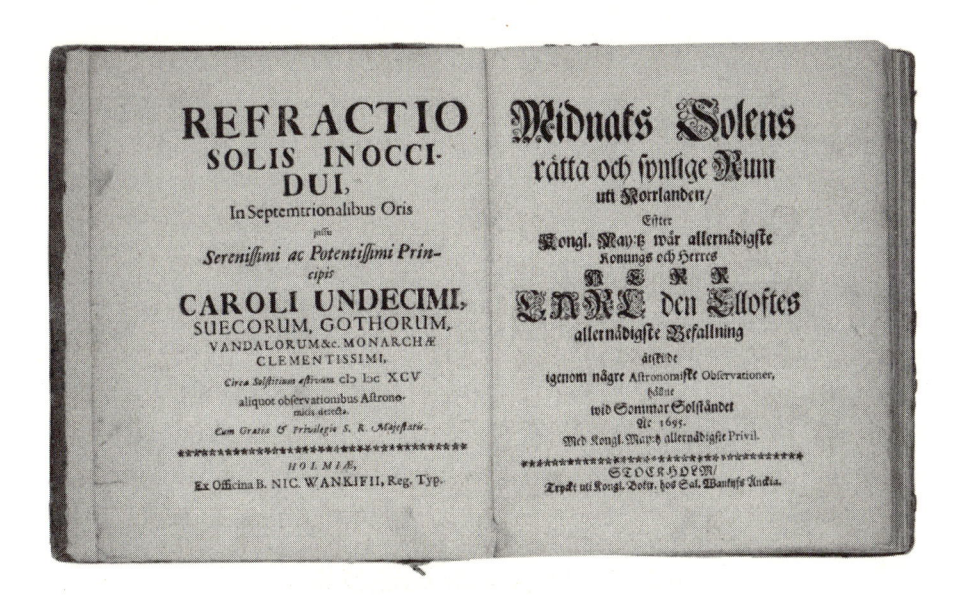

Front pages of Johannes Bilberg's account of his 1695 expedition to Lapland to study the midnight sun with Celsius' grandfather, Anders Spole.

introduction to his account of their pioneering trip to study the midnight sun and refraction of light, Bilberg firmly planted the flag of empirical science. He wrote:

> Mathematicians who have resolved never to admit of anything whatsoever without clear and perspicuous demonstration. How imperfect the knowledge of our senses is, when destitute of Reason, the Sun being elevated above the Horizon in the whole World may teach us; which yet reason dictates to be otherwise placed.[4]

This mindset also evidently left its mark on the young Celsius – guiding his lifelong approach to questioning orthodoxy and subjecting both perception and hypothesis to penetrating enquiry.

As he reached adolescence, Celsius' parents encouraged him to study law. They reasoned that such a solid and respected profession would give their surviving son much greater financial security than they had experienced. Anders was reluctant to depart from the geometry and mathematics he loved, but eventually bowed to Nils' and Gunilla's wishes and began lessons in jurisprudence. Had he continued down that path, we might never have heard of the name Celsius, let alone confer upon it such importance in our lives. But the friendship he struck up with a fellow legal student, Samuel Klingenstierna, took him in a different and history-making direction.

Klingenstierna was three years older than Celsius and came from Linköping, a smaller cathedral city about 200 kilometres south-west of Uppsala. He was a handsome, fashionably dressed and self-possessed young man with an imposing presence and twinkling gaze. Klingenstierna's father was an army officer, while his grandfathers were both bishops in the Church of Sweden. He shared Celsius' fascination with mathematics and, even before he abandoned his own law studies in favour of natural philosophy, he was confidently lecturing other students. He expounded on the advanced theories of Isaac Newton – even highlighting some conceptual errors in the Englishman's work on light refraction[5] – and Germany's master of calculus, Gottfried Wilhelm Leibniz.

Within a decade of him and Celsius first meeting each other in the legal classroom, Klingenstierna was appointed Uppsala University's professor of geometry. And he went on to become its first chair of experimental physics and later an influential figure in Sweden's royal court. His pupils included many others (such as Bengt Ferner, Mårten Strömer, Pehr Wargentin and

Celsius' fellow student and future professor of geometry, Samuel Klingenstierna (1698–1765). Lithograph by Otto Henrik Wallgren.

Jonas Meldercreutz) who would subsequently play significant roles in Celsius' life and go on to enjoy considerable academic success of their own. But these noteworthy connections all lay in the future – for now, Celsius and Klingenstierna were kindred spirits: prodigiously gifted young men, anxious to direct their talents away from the law and towards practical science.

By his sixteenth birthday, Celsius had followed his friend's example and devoted himself fully once more to mathematics. His parents were unhappy about him turning his back on a potentially lucrative legal career, but were also keen to encourage their son's enthusiasm for his chosen subject. So they arranged for him to have lessons with a private tutor, Anders Gabriel Duhre. This new teacher had himself been a child prodigy, starting his university studies aged just 14.

Despite the efforts of Petrus Elvius and others to nurture his abilities, Duhre had failed to live up to his early promise. After fifteen years' study, he left without a degree in 1710 and began teaching at the National Fortification Office in Stockholm and the Bureau of Mines. It was this latter position that, in due course, inspired and facilitated one of Celsius' first published experiments, in the depths of Sweden's biggest silver and copper mines.

Having missed out on a formal qualification in pure mathematics, and by now in his mid-30s, Duhre preferred to concentrate on its applied version. He was desperate to show how the sophisticated use of numbers could create better agricultural crops, methods and machinery. And to turn his theory into practice, he took on the Crown lease of a property called Ultuna Kungsladugård (King's Barn) just outside Uppsala, where he opened his Laboratorium Mathematico-Oeconomicum. Mathematics, Duhre believed, was not just about beautiful propositions, formulae and models, but also the hard reality of industry and commerce. Mastering the disciplines of arithmetic, geometry, topology, algebra and analysis was not enough for him; he wanted to make these skills pay their way in the modern world. And this mindset evidently rubbed off on his pupils.

The King's Barn project was a clever marriage of bucolic calm and intellectual endeavour – a uniquely Swedish salon to inspire young minds and fresh thought. It thrived for a while but eventually ended in disaster, when Duhre was outmanoeuvred by the sly Mayor of Uppsala, Johann Brauner, who persuaded Parliament to transfer the barn's lease to him. It left the visionary tutor without a home or income, and he died in poverty in 1739. But during the heyday of his novel experiment he enthused dozens of students like Celsius and Klingenstierna, igniting their interests and recommending books they should read or acquire.

Duhre had a particular flair for calculus – the mathematical study of change. He was especially fascinated by how rates of change can vary, and themselves derive from countless subsets of smaller movements. The significance of this branch of science, first explored by Greek philosophers in the fifth century BCE, can be hard to grasp in the abstract, but it comes to life when considered in more concrete terms.

A common illustration involves the halfway 'Zeno' paradox. Imagine you hold an object in your hand, which falls from your grasp. Before it reaches the ground, it must first fall half the distance from your hand, then half that distance again, then half that distance and so on. Theoretically therefore (and before the discovery of atoms and sub-atomic particles), it was possible to suggest that the dropped object would never come to rest within a finite time, although the eyes and lived experience obviously showed otherwise. In this example, the freefall object would also be accelerating throughout due to gravity, making the spatial relationships within its descent still more complex. Duhre was captivated by this field of differential calculus, using complex graphs and formulae to explore the realms of infinitely small, and purely notional yet logical, subdivisions of time and space.

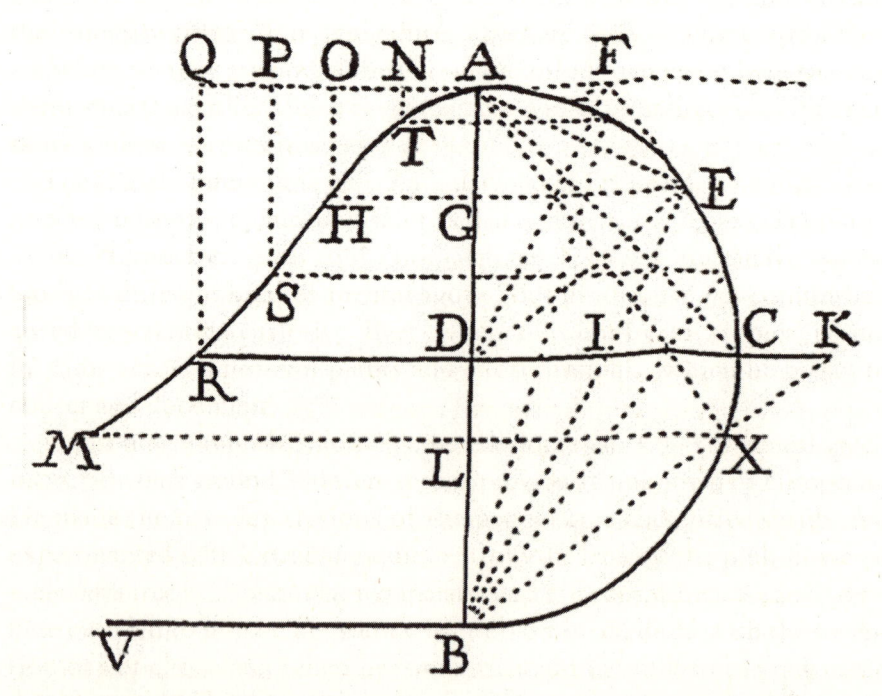

Calculations on the properties of pi (π) by Celsius' mathematics teacher, Anders Gabriel Duhre, 1721.

Whatever his shortcomings and later personal difficulties, through calculus Duhre introduced the teenage Celsius to a new and mesmerising world between reality and imagination – a liminal zone in which the student would go on to excel.

Of all the objects, documents and places I was able see while researching Celsius' life, his geometry exercise book from this time with Duhre was the most moving of all. It is kept today in the special collections of the Carolina Rediviva Library in Uppsala. Inside the book's faded marbled-paper outer cover, the owner marked and dated his name neatly in the bottom corner of the facing page. What follows is impeccably ordered and presented – each page titled with a numbered proposition and problem, explored through expansive written text and exquisitely drawn diagrams. The figures start off quite simple – triangles, circles and ellipses – but then become progressively more complicated, advancing onto elaborate curves, arcs, three-dimensional objects and rotations.

Examining the paper with a magnifying glass, I saw that, for the most ambitious drawings, Celsius had first lightly scored the paper with a stylus, then added the fine lines and annotations in iron gall ink – now brown with age, but as crisp and clear as if they had just been put onto the page. I marvelled at how anyone, least of all such a young student, could manage to execute this hair-thin precision with a feather quill pen. Several pages were lightly punctured by the point of Celsius' compass at the centre of his circles and segments, the pinholes now surrounded by a tiny ring of rust.

The whole book speaks of exceptional promise but, with knowledge of what Celsius went on to achieve in adulthood, one page stands out. While most of the book's content is laid out vertically in portrait format with immaculate margins, towards the back of the volume Celsius turned it 90 degrees into landscape. In curling capitals, the page is titled 'MENSURE LIQUIDORUM, MENSURE ARIDORUM, MENS: AGRARIAE' ('Measures of liquids, solids and agriculture'). He presents a sophisticated, three-part table, showing the relationships between different units, from Latin measures based on Roman-era drinking vessels (*amphora, cantharus, quintarius*) to the Swedish versions of a foot and mile.

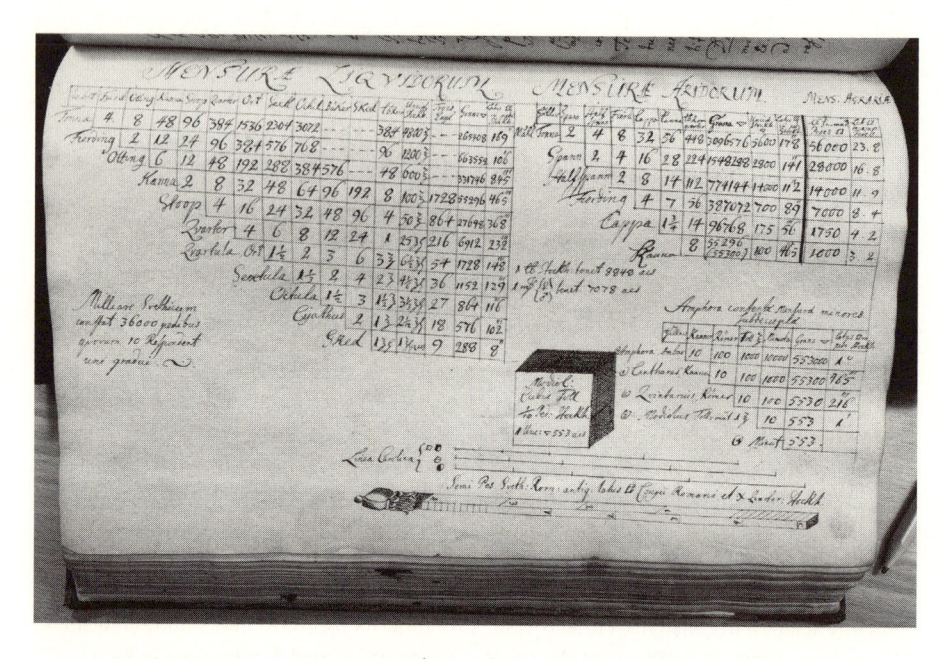

A page from the 16-year-old Celsius' geometry exercise book during the time he was being tutored by Duhre at the King's Barn near Uppsala.

This page has the air of something Celsius did for his own benefit and on his own initiative – apparently seeking to understand the connections between various units and their relationships to the real world. It also displays an innovative command of data and presentation – 150 years before the periodic table was devised and very similar in appearance to a modern spreadsheet. This is the early work of someone set to become a master of measurement, a man whose own name would become an internationally recognised unit of quantity.

There is another poignant volume from this period within the astronomy division library at Uppsala University – a textbook given to Celsius by his father Nils on St Andrew's Day, 30 November 1717. A looping inscription on the inside page reads: 'Father gave [this book] to [his] dear son Andreas Celsius, on the very day of Andreas, 1717. Nils Celsius.'[6]

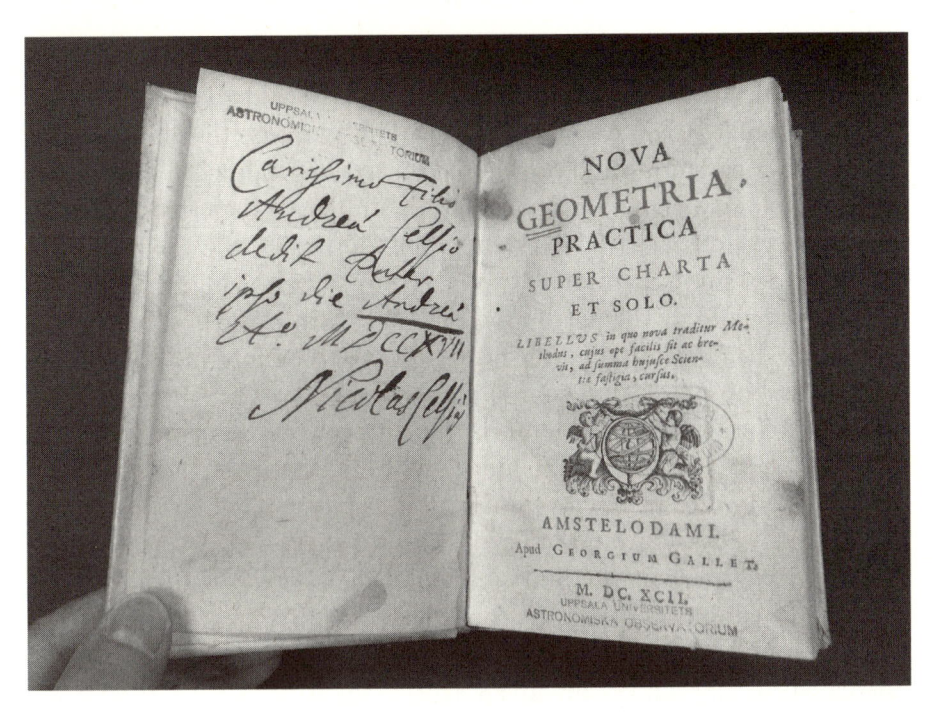

Nils Celsius' inscription recording the gift of an astronomy textbook to his 16-year-old son Anders. Photo courtesy of Eric Stempels, University of Uppsala.

From 1719, Celsius became eligible to attend astronomy lectures at Uppsala University. His first tutor was Eric Burman, the son of a parish priest from Bygdeå, a village on the Baltic Sea in the far north of Sweden. Twenty-seven-year-old Burman had himself studied mathematics at Uppsala and, after a period of teaching in Stockholm, now returned to his alma mater, where he would later succeed Nils Celsius as professor of astronomy.

On top of his academic abilities, Burman was an accomplished musician, who also took up the important position of organist at the cathedral in Uppsala. In his elastic mind, mathematics and music merged: the precise divisions of a score into beats and bars were conjoined with the repeating patterns in number theory that allowed him to calculate and predict the unknown. In this he too followed the humanist path back to classical Greece. Burman learned how it was the Ionian philosopher Pythagoras who had first sought to explain the cosmos through his 'wave theory' almost 2,000 years earlier.

Celsius' astronomy tutor Eric Burman (1692–1729), with whom he began making weather observations in Uppsala in January 1722.

Legend had it that Pythagoras happened to hear four blacksmiths simultaneously at work with their hammers and noticed that two of the hammers appeared to be 'singing the same note' with each blow. When he approached the smiths and weighed the tools, he discovered that they were 12lb and 6lb – an exact ration of 2:1, thus creating a perfect octave. And when he examined the other two hammers, he found a different relationship. They were 9lb and 8lb, which when combined with the octave hammers created further perfect ratios: 4:3, 3:2 and so on.

Pythagoras proceeded to test his theories on the strings of a lute and then develop the principles of vibrating consonance and dissonance, speculating upon their connection to the heavens – his *musica universalis* ('music of the spheres').[7] By identifying the basis of musical harmony he also revealed the mathematical foundations of the natural world:

> If earthly objects such as strings or pieces of metal make sounds when put in motion, so too must the Moon, the planets, the Sun and even the highest stars. As these heavenly objects are forever in motion, orbiting the Earth, surely they must be forever producing sound.[8]

Burman was every bit as enthralled by the patterns of rhythm, frequency, pitch and scale as his ancient Greek counterpart, and strove to pass on

An etching depicting the story of how Pythagoras (c. 570–495 BCE) developed his theories about the universe after watching blacksmiths wielding their hammers.

the wonders of these correlations through his teaching of geometry, trigonometry and calculus. And in Anders Celsius he found a willing and unusually capable pupil.

In early 1722, Burman enlisted the fast-maturing Celsius as an unpaid assistant in his latest topic of interest: weather patterns. Together, in the grounds of the University, they began to take and record daily measurements of temperature and air pressure – a regime that quickly expanded to include wind strength and direction, cloud cover, precipitation, fog and thunder. Of course, it was cold in Uppsala in February, but exactly how cold, and how would this compare to the days and years before or those yet to come? Memory and perception were unreliable witnesses; how could they obtain measures that were more objective, accurate and dependable?

In seeking to answer these questions, the two men encountered an instant obstacle – the assortment of thermometers they employed to gauge temperatures used a variety of scales (Hawksbee, Prins [later renamed Fahrenheit], Rømer, Réaumur and others), none of which seemed to correlate in a consistent way. Sometimes, even instruments calibrated to the same scale positioned right next to each other gave markedly different readings. So, in an attempt to moderate these differences, they made the times of their measurements more regular.

One or both of the men would return to the garden to record the levels up to four times every day, at sunrise, midday, early afternoon and in the evening, with the aim of capturing the minimum and maximum temperatures in each twenty-four hours. Their dedication formed the bedrock of this new discipline. But what they really lacked was a fixed standard, and high-precision tools they could apply to it with confidence.

The problem that Celsius was to eventually and famously solve had seeded itself in his mind. From these rudimentary beginnings, he and Burman created a methodical system of weather monitoring. And, with just a few gaps (notably in the following two decades coinciding with key events in Celsius' life), it has continued ever since. After marking its tercentenary in 2022, the Uppsala Series remain among the world's oldest and most meticulously kept weather records from a single location.

The relentless wave of change and invention sparked by the Enlightenment kept rolling forward. And in a frosty corner of Uppsala's university estate, Celsius took his first steps into what would eventually become climate science.

PART II

LIGHT AND AIR

☼

7

THE YOUNG PROFESSOR

Establishing his Roles and
Reputation in Uppsala (1719–32)

Youth is happy because it has the capacity to see beauty. Anyone who keeps the ability to see beauty never grows old.

Franz Kafka, *Ein Hungerkünstler* ('A Hunger Artist'), 1922

Celsius' daily weather measurements introduced him to the rigour, dedication and endurance required for serious scientific discovery. The precisely timed visits to the university garden and detailed notetaking came naturally to his already well-ordered mind, while the meteorological discipline strengthened his commitment to orderly observation. It was a springboard for Celsius' imagination that propelled the rest of his life and career.

Just as the two scientists' thermometers of various scales in the University garden gave different readings for each day's maximum and minimum temperatures, so their collection of barometers indicated a range of results for air pressure. Celsius noted not just the differences between the instruments, but also how air pressure varied by time of day, weather conditions and where he positioned the instruments. The notion that air had weight, and therefore exerted pressure on the Earth's surface, had been demonstrated eighty years before by the Italian physicist Evangelista Torricelli, who wrote: 'We live submerged at the bottom of an ocean of the element air, which by unquestioned experiments is known to have weight.'[1] The experiments Torricelli referred to were by his tutor, Galileo Galilei, and his own tests with mercury-filled glass tubes.

A few years later, another young scientist, Blaise Pascal (1623–62), studied the effect of altitude on air pressure by getting his brother-in-law to carry a barometer to the top of the extinct volcano Puy-de-Dôme in central France and then comparing the summit reading to others taken simultaneously at lower levels. Pascal was motivated by a desire to overcome the difficulties of pumping water from mines, and it seems that, in part at least, Celsius had something similar in mind. How might a better understanding of air pressure benefit Sweden's industry in silver, copper, iron, lead, zinc, sulphur and other minerals?

He decided upon a practical test, which built upon the work of Torricelli, Pascal and others. What might he discover if he took a barometer up to the highest point in Uppsala (the cathedral's towers) and also to the lowest practicable point – in the depths of silver and copper mines nearby?

Climbing the medieval cathedral tower was strenuous but straightforward enough. In December 1722, carrying a heavy, metre-long wooden and brass barometer, Celsius wound his way up the tight, spiral stone staircase into the north tower, stopping at regular intervals to catch his breath and record the level of mercury in the tube. And then he did the same in reverse, gingerly descending the narrow steps – more hazardous in this direction – and noting the results alongside his upward figures. After repeating this several times, Celsius wrote in the front page of an almanac he'd just bought: 'From the observations I have hitherto performed in the

An eighteenth-century view of Uppsala showing the cathedral tower Celsius climbed with his barometer. Lithograph by Auguste Etienne François Mayer.

tower of the Uppsala cathedral, I have found that for one line of mercury falling in the barometer there corresponds 96 feet in height.'

The ascent of the cathedral tower confirmed the findings of Torricelli and Pascal – that air pressure reduced with altitude – and took Celsius into the realm of his first original scientific observation. He had quantified how much and how consistently this change occurred. Now it was time for him to examine the inverse – how did air pressure *increase* with depth, and what was its relationship to air temperature?

Sala lies about 55 kilometres west of Uppsala – a settlement that grew up around the town's silver mine. By the time Celsius visited in 1724, its deep shafts had already been producing precious ore for more than two centuries. Much of Sweden's coinage had originally been cast from Sala silver, and it had also been blended with metal from other mines to create the ornate coronation throne, which sat in the royal apartments in Stockholm, ready for the next monarch.

The metamorphic bedrock at Sala is almost 2 billion years old – laid down during the Paleoproterozoic era, as Earth's continents began to form and stabilise. Crushed and suffused by intense heat and pressure, the minerals inside Sala marble give it a distinctive, pale green tinge – a clue to the wealth the stone can yield. But Celsius came in search of a different treasure: truth and its practical application.

☼

Three centuries later, my own visit to Sala revealed the acute dangers and conditions the 22-year-old professor encountered in his quest. On a sunny June morning, I met the mine's charismatic historian, Niklas Ulfvebrand, at one of the few buildings still surviving from the eighteenth century. This was the administrative centre where Celsius most likely began his exploration. He joined the mine manager and around 150 workers massed at 5 a.m. for morning prayers to give thanks for the silver and ask God for safety, before being winched down into the 200-metre-deep darkness of the Queen Christina shaft.

Nothing in Celsius' life and studies up to this point could have prepared him for such an experience. He was an astronomer – more used to gazing up at the stars and planets above than penetrating the Earth's surface. The shift supervisor would have given Celsius a pair of crude wooden pattens to protect his feet and provide some grip underground, and one of the drooping felt hats the miners wore to deflect flying splinters of rock away from their eyes. Niklas showed me a *tunna* – a loose barrel of thick

An eighteenth-century schematic of the sprawling Sala silver mine – its tunnels were 1 kilometre deep and 20 kilometres long. (Image courtesy of Sala Silvergruva)

staves in which everyone had to be lowered – miners, managers, scientists, even royalty when they came to inspect the natural riches of their kingdom. Lives depended on the skill of the men operating the water wheel, which powered this rudimentary lift via a 30-metre-long series of clanking wooden links. The slightest error in adjusting the flow of water onto the wheel or a break in any of the couplings could mean a fatal jolt and plummet to the bottom.

Those, like Celsius, not yet hardened to the fear and vertigo, tended to crouch in the centre of the *tunna*, surrounded by the miners' tools and supplies. He would have held his precious barometer and looked up at the apparently fearless men nonchalantly sitting or standing on the edge of the barrel. With one hand they held onto the chain that connected them to the winch, and in the other they each held a *torrvedsbloss* – a torch made of slow-burning dried pine splints rammed together into a tight iron ring. By this flickering flame, the only source of light underground, Celsius would have seen the mercury creep up the tube, and feel the pressure building in his ears.

My journey into the mine was much less perilous – a gradual descent by airy but safe stairs, with solid platforms and walkways now spanning chasms that in the eighteenth century would have been completely open and unprotected. As Niklas and I took an electric lift down to the lowest point above the water table, I too felt my ear drums compress and watched the black numbers on my digital barometer steadily climb – from 1012 hPa at the surface to 1031 hPa at around 155 metres. The measurements on my modern handheld device were in hectopascals (1 hPa = 100 pascals) – another international unit taken from the name of a scientist who pioneered this field of exploration.

In the days before radios or telephones, communication within the mine and up to the surface was governed by a strict vocal protocol. Before he entered, Celsius would have been drilled in the rules: no whistling, screaming or cursing. These proscriptions allowed the all-important commands of '*Ropare!*' (rope/lower) and '*Skjuv upp!*' (pull up) to be shouted and clearly heard up and down the shaft.

The miners were however permitted, in fact encouraged, to sing – the Lutheran hymn '*Vår Gud är oss en väldig borg*' ('A mighty fortress is our God') often echoing around the mine as shifts began. To bolster the miners' courage, one verse proclaimed:

> And though this world, with devils filled,
> should threaten to undo us,
> we will not fear, for God has willed,
> his truth to triumph through us.

Sala's silver miners descending in a *tunna*, by the light of a *torrvedsbloss* torch. Painting by Bo Svärd, 2013.

Niklas treated me to a beautiful rendition of the hymn when we paused around the midpoint of the Queen Christina shaft. As his voice soared up the cavernous holes bored into the rock, I glanced back at my barometer – the reading remarkably consistent with Celsius' estimate that each line of mercury in his tube corresponded to 106 feet of depth (equivalent to 1 hPa per 10 metres).

As we splashed through some of the narrower tunnels deeper down, I put my hand into a crevice to steady myself and felt something smooth and soft. Shining the torch onto my fingers I saw a wet, silvery paste. 'Magnesium silicate,' explained Niklas, a natural source of the talcum once used in cosmetics and baby powder. Even now, the mine is still forming and giving up its minerals.

The vast human effort involved in mining here really struck home when Niklas showed me the evidence of the *tillmakning* (firesetting) process used to create the shafts and release the valuable ore-laden stone. Workers would light fires at the rock face, leave them to smoulder overnight then extinguish them the next morning. Once the rock had cooled it was brittle enough to be hacked out and transported up to the surface by men working in relays, with cradles and handbarrows.

I asked Niklas how much progress they could make in a day using this method. He put his hand against the mine wall and opened his forefinger and thumb across a faint stripe in the rock – about 12 centimetres wide. At this rate, each metre of excavation took a month, and the Sala mine is a kilometre deep with 20 kilometres of passages and caverns, some as big as a church. On average, the miners discovered viable deposits in about half the tunnels they dug – a 50/50 chance of hitting the jackpot, or a lot of wasted effort. Depending how you look at it, these were odds worth taking, or a marginal bet on the potential reward. Either way, the miners worked ten- or twelve-hour shifts, six days a week. In winter, most of them only saw daylight on Sundays.

Niklas also demonstrated how Sala's ingenious engineers overcame the problem of pumping water, which had sparked Torricelli's fascination with air pressure in Italy the century before. The water was brought up through a ladder of thirty-two hollowed-out tree trunks, each feeding into the one above it. A tight-fitting piston in each section was powered by horses kept underground turning a heavy, geared wheel. This raised the water by the maximum gravity would allow – around 9 metres – before filling the next sump to be propelled up to the following tube, with more gears and horses, and so on. Without this technology, Sala could never have produced such riches. And if the pumps had stopped working, within a couple of years, the mine would have steadily refilled up to the natural water table just 16 metres below the surface.

☼

Celsius' visit to Sala showed him how air pressure increased with depth. But would the rate be the same everywhere, or did it depend upon the surrounding soil and rocks? Again he resolved to find out for himself, and headed north to Falun – a five-day journey to the huge copper mine there, from which the distinctive red *falu-rödfärg* paint emerged as a by-product to decorate and protect the exterior of Swedish houses and barns. This mine is even bigger and older than Sala – a mountain that yielded up its elemental treasure for 1,000 years. It is now a UNESCO World Heritage Site.

When Celsius visited, the mine was nearing the end of its heyday, but still producing most of Europe's copper and making a sizeable contribution to the Swedish Exchequer through the same time- and labour-intensive firesetting method used in Sala. One contemporary account declared it was 'one of the great wonders of Sweden, but as horrible as hell itself', with the miners surrounded by 'soot and darkness on all sides. Stones, gravel, corrosive vitriol, drips, smoke, fumes, heat, dust, everywhere.'[2] All around the mine, the spoil was tipped into enormous open hearths and alternately roasted and smelted to produce the raw copper – a process that could take up to a year to produce a single, exportable ingot. Long before Celsius reached Falun, he would have seen the soaring columns of smoke from these fires and caught their unpleasant sulphurous whiff on the wind.

On Midsummer's Eve 1687, the riven ground around the mine began to emit loud groans – the plaintive sound of a natural resource pushed beyond its limits. With the workings below luckily empty of personnel because of the summer holiday, the whole mass spectacularly imploded, leaving the 100-metre-deep chasm that still exists today – an open, red wound attesting to the folly of depleting the Earth's reserves to the point of collapse. Commercial mining continued here until 1992, but never at the same rate, and it was progressively overtaken by the vast deposits and industrial extraction in Asia, South America and Australia.

In the deepest Falun mineshaft, Celsius' thermometer indicated that, allowing for the air temperature within the mine, the relationship between air pressure and depth was 1 hPa per 8 metres – close to the exact ratio now measured by modern equipment. The explanation seems obvious now, but was unproven at the time: that the weight of all the molecules in the Earth's atmosphere press down upon the planet's surface. At sea level, every square metre is exposed to around 1,000 hPa of pressure – roughly half what you put into the tyres of a car. But the higher you go, the less atmosphere there is above, and hence less pressure. Conversely, the deeper you go below the surface, the more intense the weight and temperature become.

Storytellers have long fantasised about drilling deep towards the centre of the Earth, but we now know that the rate of pressure increase alone is so rapid as to make any form life impossible more than about 10 kilometres down – barely a third of the way into the planet's outermost crust. Humanity is destined to exist only upon the thin exterior of its home.

Eager to share the insights he had gained from so much exertion, Celsius submitted a short paper to the Swedish Royal Society of Sciences, which was published in late 1724. The two pages recorded his ascent of the cathedral tower and the visit to Sala on 28 August that year, in each case noting the accompanying weather conditions: 'In the range of the hour, during which the whole observation was accomplished, the sky was somewhat rainy and windy; yet no sensible variation was visible during this time on the mercury column in another barometer attached to the wall above the mine.'[3]

To Celsius' great delight, the Royal Society in London – an institution where he would eventually become a member – also included his findings in its digest of *Philosophical Transactions*. The slim paper's findings were not just a matter of scientific principle; Celsius and the two societies clearly saw the scope for practical and commercial application of this new understanding about air pressure. Analysing and predicting the atmospheric force at different heights, depths and locations was useful not just for mining, but also made it possible to chart the topography of land more accurately. Combining barometric readings with trigonometry pinpointed peaks, valleys and contours that could be relied upon for farming, forestry and industry.

The rise and fall of mercury in Celsius' glass tube also proved something fundamental beyond all doubt. Since the barometer's behaviour was consistent and predictable, it followed that the Earth's atmosphere must be finite. And compared to the already known vastness of the solar system and space beyond, it must also be a relatively thin layer. At the age of 23, Celsius had acquired both potent new knowledge and international attention.

But just as his adulthood and career began to blossom, his father's ended. Nils died on 21 March 1724, aged 65. His energies and self-belief had never recovered from the rough treatment he'd received at the hands of the Church of Sweden and the opprobrium heaped upon him in response to his daring student dissertation forty-five years before. In later life, Nils had cut rather a sad and diminished figure, resentful perhaps of his more successful brothers, embarrassed maybe by needing to rely on his wife's and daughter's eating house income to support the family, and hurt at not being able to fulfil his own potential.

What Anders Celsius thought and felt at the time of his father's death was either not recorded or has not survived. But the vigour with which he pursued his early studies and took on ever more responsibility suggests that it spurred him on. He was energised, hungry and determined not to suffer a similar fate to Nils.

☼

The published reports of Celsius' air pressure experiments opened the door for him to work as an unpaid assistant to Sweden's Royal Society of Sciences. And just a year later he became its secretary, with particular onus on him to produce the society's publications. The post was still without a salary though, so he continued to cast around for other roles that might bring him some financial reward.

Two openings arose in 1728. First, his old student friend Klingenstierna was appointed by the University Senate as professor of mathematics. But this award was made *in absentia*, since he had already departed for an educational tour of Europe, studying advanced mathematics in Germany, Switzerland, France and England. Celsius volunteered to stand in for his colleague, so was made deputy professor, but still with no pay.

Desperate for money, he reached out to his old teacher Duhre, who invited him to come and work at the King's Barn just outside the city. Celsius gratefully accepted the offer and began lecturing to sizeable classes there, on top of his university duties. Somehow, he also found the time to write an elementary textbook to accompany his courses,[4] and he proved himself a patient and capable tutor.

Meanwhile, Celsius continued with his own studies, by presenting two theses. With Burman as his astronomy supervisor, he wrote '*Disputatio astronomy ca de motu vertiginis lunae*' ('An Astronomical Lecture on the Rotational Motion of the Moon'). And for philosophy, the professor of law, philology and politics, Johan Hermansson examined him on a fifteen-page tract '*Dissertatio gradualis de existentia mentis*' ('Observations on grades of mental existence').[5]

In this second thesis, Celsius dug deep into philosophical foundations, following Descartes' notion of 'I think therefore I am' to establish the reality of existence and existence of reality. A sentence in the preface

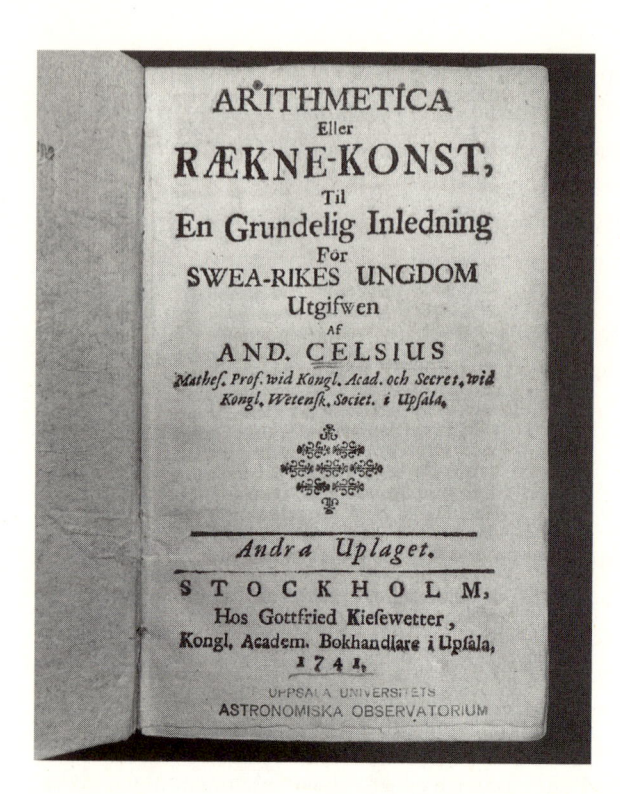

The textbook Celsius wrote for his young students in 1728: A thorough introduction to arithmetic and numeracy for Swedish youth.

signals his fast-developing rational convictions: 'It is a duty of a philosopher not only to know what could happen and what could not, but also to have clear in mind the reasons why something could happen, or could not be.'

Both presentations were well received, demonstrating Celsius' growing confidence and command of his subjects. But with so much on his shoulders, the pressure was beginning to tell. In 1729 he wrote to the Swedish Royal Society's founder, and former university librarian, Eric Benzelius: 'I hope that when Prof. Klingenstierna returns home, which is said to occur this autumn, he will take away half my burden.' But by November that year there still was no sign of Klingenstierna, and Celsius lost another important person in his life.

Burman died suddenly. He was just 37 years old, but with his furrowed brow and thick grey beard and moustache, had looked much older. The colleague with whom Celsius had begun the pioneering weather measurements, and who had also served as the cathedral's organist and director of music, was gone. Within five years, while Celsius' own capabilities had started to burgeon, he had been deprived of his two biggest intellectual influences: his beloved father, and now his sharp-minded tutor and collaborator. But, as before, this unexpected change brought opportunity: the University needed a new professor of astronomy.

So it was that Celsius took on yet another job – his fifth – initially as deputy professor. Crucially though, this came with a modest salary, and he was able to press his case for taking on the full professorship. With Burman gone, Celsius assumed lead responsibility for recording the Uppsala weather observations. Each month, he faithfully produced tables of the daily temperature and air pressure data, along with accompanying comments about any notable meteorological events or conditions.

With this established activity ticking along in the background, Celsius polished his credentials to become the new professor. He pledged that he too would undertake a scientific tour of Europe, with the specific aim of visiting the continent's greatest observatories to bring back a realistic concept for Uppsala to have its own, purpose-built laboratory of the heavens. Celsius also pointed out that, while working on submissions to the Royal Society in London, he had learned to read, write and converse quite well in both French and English – further important additions to his fluency in Latin and other skills.

When the Senate met in its gilded meeting room opposite the cathedral to review the five applications for the vacant astronomy position, Celsius was the unanimous favourite. He had the family heritage, the depth of learning, the vision and, above all, the desire to take his science into new and exciting territory. He was the right person for the job – as long as the King, who was still the final arbiter over senior university appointments, agreed.

Royal approval eventually arrived in early summer 1730. Anders Celsius became the third successive generation of his family to hold the title of professor of astronomy in Uppsala. Secure at last, he relinquished his teaching with Duhre and, on 2 June 1730, delivered an inaugural lecture, '*Astronomiae usus in vita civili*' ('The use of astronomy in civil life'). The new professor had arrived and clearly wanted to demonstrate the down-to-earth utility of his science. But he would have to wait for Klingenstierna to return before he could embark on his own travels.

For two more years, Celsius busied himself with teaching and research. He gave compelling lectures on spherical trigonometry – using a spherical chalkboard to show his students how to measure the curved triangles of land or the night sky for accurate surveys, observations and navigation. He also worked on measuring the latitude of Uppsala by the Sun, began exploring Baltic Sea levels and made further inroads into studies of temperature and air pressure. And even when Klingenstierna finally reappeared from his tour, there was one more thing Celsius wanted to do – to help out a new friend.

☼

Celsius' uncle Olof, brother of the late Nils, was Dean of Uppsala Cathedral, an enthusiastic amateur botanist and a master of languages. From his father Magnus he had inherited a particular fascination with Scandinavia's rune stones and carvings. Although the botanical garden in Uppsala, originally created by Rudbeck the Elder, had become overgrown and neglected since the fire of 1702, it remained a tranquil place, where Olof liked to stroll and think. One day in 1728, he came across a young man in the garden whom he did not recognise. The stranger was polite and assured, but poorly dressed and obviously in need of a good meal. He introduced himself as Carl Linnaeus, 21 years old, the son of a parish curate in southern Sweden. After attending university in Lund, he had just arrived in Uppsala, hoping to study botany and medicine.

The two men struck up an instant rapport and fell into a lively discussion about their shared passion for plants. Linnaeus explained that he was desperate to follow his interest in the natural world, but in a new and

Celsius' uncle Olof (1670–1756).

unfamiliar city he had no money for food, clothes or accommodation. Olof took pity on the young man and invited him to come and stay with his family and use his own extensive library until he got settled as a student. By way of informal mentoring and support, Olof also offered to put Linnaeus in touch with his astronomer nephew Anders, who was just a few years older and knew the University well. The two would get along well and help each other out, he thought.

In his life and education up to this point, Linnaeus had been generously supported by a number of benefactors: priests, academics and others who recognised his exceptional intellect and promise. Through this chance encounter with Olof Celsius and his extended family, he received another welcome leg-up: a direct link to one of Uppsala's notable academic clans. And in Anders Celsius he found a loyal confidant and soulmate – a cerebral equal, someone with just a little more life experience – and now, as a full professor, a wise head who could guide him in the city and its ways. From this point on, their names and reputations became intertwined.

Shortly after Celsius had delivered his second thesis, Linnaeus produced his own. '*Praeludia Sponsaliorum Plantarum*' ('On the prelude to the wedding of plants') was written in a curling hand and illustrated with naive drawings of plants casting their pollen to the wind and being nourished by sunshine and rain.[6] His focus was on how plants sexually reproduced. This was not an original insight, but was a subject on which Linnaeus would soon become the acknowledged expert and spokesman.

Like Nils Celsius' dissertation half a century before, Linnaeus' writing attracted outrage. Some of its early readers were indignant at the mere notion of God's most beautiful and delicate floral creations being involved in anything as sordid as sex. Some angrily accused Linnaeus of blasphemy – but not Olof Celsius. He showed the thesis to Olof Rudbeck the Younger. Rudbeck was impressed by the clarity and force of Linnaeus' argument and asked him to become 'demonstrator' of the botanical garden.

The role was a combination of scientific overseer, writer in residence and public relations manager. The young man's on-site lectures quickly attracted attention, with audience numbers rising from a few dozen to 300–400. People were so anxious to hear Linnaeus' theories about nature that they spilled out of the small, first-floor lecture room to surround the house in the corner of the garden. Linnaeus spoke in Latin, but with the trilling accent of his Småländska home region, keeping his audiences spellbound with a mix of humour, narrative and revolutionary thought. He was a persuasive, natural-born communicator and educator.

Carl Linnaeus – Celsius' young botanist friend, and the future father of taxonomy. Statue by Carl Eldh.

Frontispiece of Carl Linnaeus' handwritten and self-illustrated 1729 thesis on sexual reproduction by plants.

In consort with Olof Celsius, Linnaeus continued to develop his theories. The two men made several outings together into the surrounding countryside, studying and collecting plants including the tiny woodland species later named *linnaea borealis*. The twin, bell-shaped, pink and white flowers of this delicate shrub presaged and became the symbol of the binomial system of taxonomy that Linnaeus would go on to establish – designating every living thing with just two Latin words.[7]

But that lay ahead; for now he moved into the house of Rudbeck the Younger, where he taught some of the family's younger children and heard tales of his host accompanying Anders Spole and Johannes Bilberg to Lapland in 1695. The detailed records of that expedition had been lost in the 1702 fire, so perhaps, thought Linnaeus, it was time for a repeat visit.

He was sure he could discover new plants, animals and perhaps even valuable minerals in this inhospitable and sparsely populated region. He also wanted to know more about the lifestyle, customs and culture of the Sámi, the tough nomads who roamed the Arctic tundra with their reindeer herds. Linnaeus applied to the Swedish Royal Society of Sciences for a grant to support such a trip – his request supported by Olof Celsius and processed by the society's secretary, Anders Celsius. With their influence behind him and Linnaeus' growing reputation for communicating practical science, it was approved.

To help him prepare and plan the expedition, Linnaeus enlisted Celsius' support. As a child, he too had heard stories of his late grandfather Spole's encounters with strange landscapes, customs, flora and fauna in the far north. So he was enthused about the idea of his popular botanist colleague undertaking a similar adventure. Together, they pored over maps, plotted routes and assembled the equipment and provisions that Linnaeus would need. In doing so, Celsius was both reaching back into his family's store of knowledge and unconsciously forming the template for his own Arctic journey in years to come.

☼

Bolstered by Celsius' exhaustive preparations, at eleven o'clock in the morning on Friday, 12 May 1732, the eve of his twenty-fifth birthday, Linnaeus set out alone and headed north into the Uppland countryside. The first two pages of his journal paint a vivid picture of how he looked and what he saw – the perfect example of his storytelling art:[8]

> My clothes consisted of a light coat of Westgothland linsey-woolsey cloth without folds, lined with red shalloon, having small cuffs and collar of

shag; leather breeches; a round wig; a green leather cap, and a pair of half boots. I carried a small leather bag, half an ell in length, but somewhat less in breadth, furnished on one side with hooks and eyes, so that it could be opened and shut at pleasure. This bag contained one shirt; two pair of false sleeves; two half shirts; an inkstand, pencase, microscope, and spying-glas; a gauze cap to protect me occasionally from the gnats; a comb; my journal, and a parcel of paper stitched together for drying plants.

I wore a hanger at my side, and carried a small fowling-piece, as well as an octangular stick, graduated for the purpose of measuring. My pocket-book contained a passport from the Governor of Upsal, and a recommendation from the Academy.

At this season Nature wore her most cheerful and delightful aspect, and Flora celebrated her nuptials with Phoebus.

> *Omnia vere vigent et veris tempore florent*
> *Et totus fervet Veneris dulcedine mundus.*
> [Spring clothes the fields and decks the flowery grove
> And all creation glows with life and love.]

Carl Linnaeus depicted in local costume during his six-month expedition to Lapland in May to October 1732.

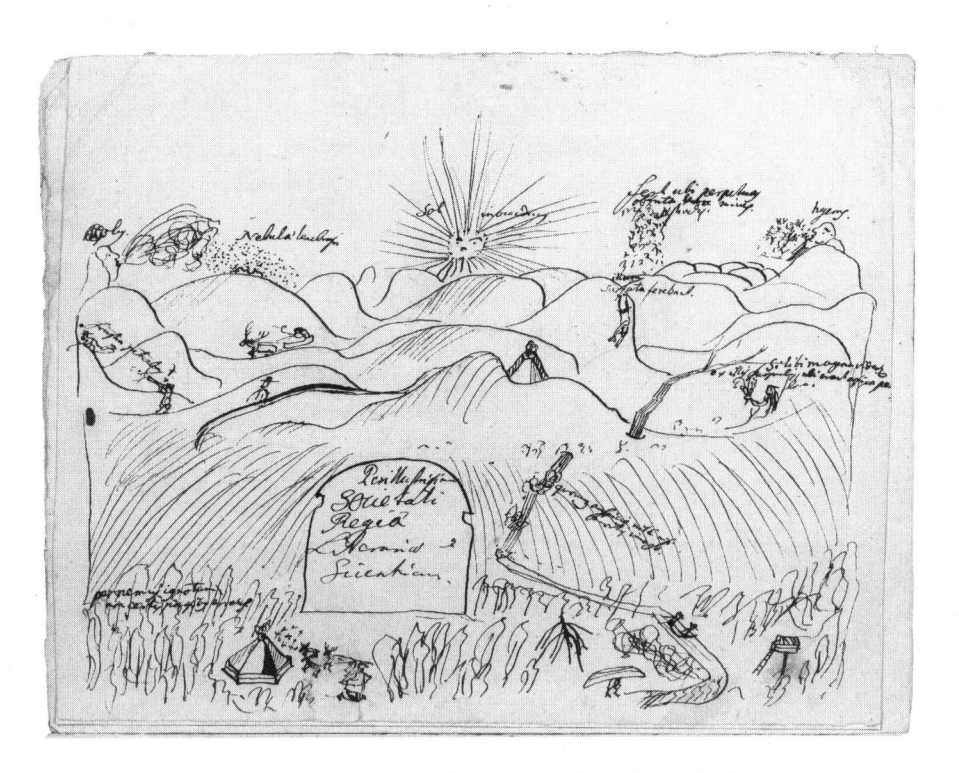

Linnaeus' sketch of various episodes on his journey through Lapland. (Image courtesy of the Linnean Society)

> Now the winter corn was half a foot in height, and the barley had just shot out its blade. The birch, the elm, and the aspen tree began to put forth their leaves.

Linnaeus' exuberance would soon be tempered by the punishing reality of a 2,000-kilometre overland trip by horse and on foot. He was tormented by insects and shocked by the primitive and – from his viewpoint – almost masochistic living conditions of the Laps. Linnaeus took a decidedly colonial view of the communities he passed through, remarking on the squalid tents in which the herders lived, where smoke rendered many of them blind. He suggested in his diary that the natives should receive 'fifteen lashes apiece' until they learned how to build proper chimneys. Most of all, though, the young traveller was troubled by the suffering of reindeer from distemper, which he recorded caused them to 'die with horrid bellowing'.

Linnaeus was nonetheless enchanted by the virgin forests and majestic grasslands, and the otherwise healthy habits and lives of the local population. In these more positive moments, he was moved to further colourful prose: 'The tranquil existence of the Laplanders answers to Ovid's description of the golden age, and to the pastoral state as depicted by Virgil. It recalls the patriarchal life, and the poetical descriptions of the Elysian Fields.'

Whatever the judgements and hardships along the way, Linnaeus' account of his trip sparkles with the sheer joy of travel as learning, and the catalytic effect it had upon the big ideas swirling around in his mind. He identified and logged thousands of plant and animal species, forming the knowledge base for his later work on taxonomy.

Meanwhile, in Uppsala, now mature and finally secure, it was Celsius' turn to follow and seek out the same fulfilment in other countries.

8

CELSIUS' GRAND TOUR

Learning from Europe's
Great Astronomers and Observatories

A man has not fully lived until he experiences that gentle balmy clime of ancient empires, the land of lemon trees and the genius of Michelangelo.

E.A. Bucchianeri, *Vocation of a Gadfly*, 2018

It is commonly held that travel broadens the mind. Another proverb says that good things come to those who wait. Both were true of Celsius' four-year Grand Tour; a much-delayed and long-anticipated adventure that sent him on a 5,000-kilometre journey through the heart of Europe, to some of its finest cities, tallest mountains and most captivating centres of thought.[1] It catapulted the young professor from quiet Nordic promise to a place on the highest-profile scientific expedition of his age.

The concept and form of the Grand Tour originated in Britain in the mid-1600s. It was a novel way for young *milords* to finish their education, build their character and prepare for life as an aristocrat, diplomat or other member of the elite. It centred on visiting and immersing the young travellers (most of whom had been drilled in Latin from an early age) in the classical world and the High Renaissance cultural and artistic riches of France and Italy. Paris, Venice and Rome were the essential stops on every itinerary, with Geneva, Lausanne, Leiden, Heidelberg, Turin, Milan, Florence, Pisa and Naples also much frequented.

For academics, these tours took on serious scholarly and scientific import as well as personal significance. They offered cultivated minds, fresh from university, opportunities to meet and work alongside their international peers and spend time at other important places of learning. In Celsius' case,

Europe's famous observatories were the principal focus; he wanted to see how these institutes operated and to assess the quality of their instruments. The plan was for him to then return to Uppsala and create an observatory there. As always, his explorations had a practical purpose.

☼

The logic and potential of the Grand Tour were unanswerable. It was designed to send the next generation of leaders, thinkers and decision-makers out into the world, to boost their confidence and test their wits through exposure to the unfamiliar. In a literal sense, it aimed to extend the tourists' horizons through direct engagement with diverse geography, language, architecture, archaeology and cutting-edge study. It became the indispensable path to intellectual self-improvement for Europe's 20-somethings, the epitome of *utile dulci* – blending the useful with the agreeable. The Grand Tour was an ostentatious symbol of advantage, but one with enlightened intentions to bolster patrician responsibility, empathy and self-assurance. The families who despatched their privileged sons and daughters in this way hoped they would return with renewed independence, emotional stability, openness and compassion. The Grand Tour was a prototype gap year of magnificent proportions.

But there were risks too for the intrepid intelligentsia. In the late seventeenth century, even crossing the English Channel was a perilous undertaking. And the rough and unsurveilled routes across thinly populated and still largely rural states were hunting grounds for violent highwaymen and brigands. For these reasons, while the richest travellers often took with them a lavish retinue of servants, valets, cooks and coachmen, they rarely carried large quantities of money or treasure. Letters of reference and introduction to royalty and high society were their door-opening currency, backed up by credit notes to redeem with banks and financiers along the way.

These downsides and challenges were part of the package – to confront young men and women of status with novel sights, sounds, smells, tastes and sensations to spark synapses, develop insights and sharpen every instinct. Once they embarked on their tours, no one would tell them what to do, what to think or how to behave – they would have to survive and thrive by meeting new people, reinventing themselves (if they wished) and hopefully becoming more aware of their own habits and capabilities.

The final keynote of the Grand Tour – usually unspoken, but widely understood and encouraged, for male travellers at least – was sex. Releasing hormonally charged young adults from the strict protocols of high-class living to explore accountability-free physical pleasures away from the threat of scandal was another important facet. It was a tacit but universal understanding that gentlemen would 'come of age' in every respect, as they discovered the eye-catching attractions of other countries and climates.

The Grand Tour continued and spread through the eighteenth century to include the prosperous youth from most of Protestant Europe. Young travellers from Scandinavia, Germany, the Netherlands and even some from the newly formed United States of America joined their British counterparts. And, towards its end, larger numbers of liberated and free-thinking women also began to enjoy its benefits.

The 1789 French Revolution brought all this to a halt. It was no longer safe or wise for inexperienced members of the nobility to range across Europe, flaunting their affluence and indulging their whims. And in the nineteenth century, when the spread of rail travel meant that longer-range leisure and excursion came within the means of more ordinary folk, the Grand Tour's exclusive appeal rapidly diminished.

The century and a half during which Grand Tourists blazed a bright trail through Europe melded the ruling classes of disparate nations together in a way that still resonates with the boundaries, alliances and norms of today. It spawned cooperation, marriages and understanding, which, despite the wars and convulsions of the period that followed, soothed deep-seated rivalries, and – for a while at least – reduced cross-continent conflict.

The Arrival of a Young Traveller and His Suite During the Carnival by David Allan, 1775. (Photo by permission of the Royal Collection Trust)

The adventures of these European graduates and nobles also created the foundations of another world-changing phenomenon: tourism.

A drawing by the English artist David Allan (1744–96) in the British Royal Collection depicts the arrival of a young traveller and his entourage in Rome's Piazza di Spagna at carnival time. His carriage is surrounded by a swarm of local hustlers, touts, street performers, pimps and prostitutes, all keen to relieve the unwitting visitor of his money and morals.

At its best, the Grand Tour exchanged cross-border goodwill, learning and cooperation to a previously unseen extent. The fashion for commissioning time-bending portraits of the travellers in imaginary classical settings or amid sweeping city views and landscapes was indulged by Byron, Goethe and many others. Their orders launched and sustained the careers of some of the great painters of the age, such as Canaletto and Panini.

For serious scholars like Celsius, independent travel through Europe unleashed potential, cemented reputations and helped to earn the brightest minds their places in history. Those who approached the undertaking with diligence returned older, wiser and ready to share their knowledge and experiences with others – contributing to the common intellectual capital of the Enlightenment.

At its worst, though, the Grand Tour was a flamboyant affectation that taught its participants little. The ships and prospects of the most unfortunate foundered on the rocks of Normandy or Brittany – fatally thrown off course by unfavourable winds and violent seas. Some young lords returned laden with little except gaudy souvenir acquisitions from their trip: paintings, crafts and statues of dubious quality or antiquity. Too many came home infected with venereal disease, their journals and minds still largely empty. Generations of unwed continental mothers and fatherless children were also forgotten and left to fend for themselves in the wake of unscrupulous visitors.

The attitudes and behaviour of these seventeenth- and eighteenth-century tourists reflected another unwholesome face of humanity that still reverberates. The advantaged young people venturing south from Great Britain, the Low Countries and Scandinavia often carried with them a strong sense (unconscious or otherwise) of northern racial, ethnic and religious superiority. Every step of the Grand Tour was accompanied by the mindset that legitimised slavery and imperialist abuse.

Paris, where classes in dancing, fencing and riding were de rigeur, was widely seen as the premier destination. And, at the time Celsius set sail, volcanoes were the focus of faddish attention and fascination. Travellers ventured to the smouldering rims of Mounts Vesuvius, Stromboli and Etna to peer into a primal underworld so different to the pristine comfort of their normal lives.

But alongside these fashionable and natural attractions, it was commonplace for visitors to view the ancient ruins and rustic ways of Italy as quaint, backward or iniquitous – ripe for ridicule or for their artefacts to be taken away. The darker-skinned women of the Mediterranean were also frequently treated with a similar lack of appreciation. They were assumed to be somehow 'naturally' looser in their morals, with untamed, primitive impulses that were fair game for the more lascivious of the Grand Tourists.

☼

True to his upbringing, education and thirst for knowledge, Celsius' account of his own Grand Tour indicates that, for the most part, he pursued the higher ideals of the tradition. He moved as quickly as the rudimentary roads of the time allowed, while stopping and staying for as long as he could at the most significant universities and observatories. After years of subsisting on his modest earnings from lecturing and tutoring, he was far from rich. So even if the more extravagant pleasures of the Grand Tour had appealed to him, it is unlikely he could have afforded them, and certainly not to the excess demonstrated by some of his contemporaries.

It was normal for most Grand Tourists to be accompanied by a chaperone or *cicerone* – someone of greater maturity and worldliness (often appointed by the traveller's family as a discreet, trusted presence) to guide the innocents abroad. This was the role that Celsius' grandfather Anders Spole had performed for the teenage sons of his army general employer. With reference to the roaming entertainers of the day and their unfortunate animals, these escorts were mockingly known as 'bear-leaders'. They acted as companions, tutors, guides and guardians, opening doors to royal and blue-blooded households across Europe, and hopefully keeping their charges out of trouble.

Celsius' choice of companions (both of whom came from his home city of Uppsala) says a lot about his motivations, intentions and means. Aged 31 by the time he began his tour, he was accompanied only by the cartographer, surveyor and engraver Georg Biurman (just one year older) and one of his brightest mathematics students, Jonas Meldercreutz, who was barely 17. As well as a statement of economy and purpose, the trio suggests a maturity and confidence far beyond their years.

Adventure and opportunity lay ahead for them all.

Goethe in the Roman Campagna by Johan Heinrich Wilhelm Tischbein, 1787.

Caricature by Pier Leone Ghezzi of an early eighteenth-century 'bear leader' escorting a young Grand Tourist.

One of Celsius' young Grand Tour companions, Jonas Meldercreutz. He went on to become professor of mathematics at Uppsala University in 1751.

9

TWO COUNTRIES, FOUR SISTERS

Travels and Studies in Germany and Italy (1732–34)

In the end we retain from our studies only that which we practically apply.
Johann Wolfgang Von Goethe, from *Conversations with Goethe in the Last Years of His Life* by Johann Peter Eckermann, 1839

Celsius' journal from his travels through Europe survives in the Carolina Rediviva Library at Uppsala University. Inside its beaten, tan leather cover are nearly 200 pages of jottings, transcriptions, sketches, diagrams, calculations and meticulous observations, with text constantly jumping between Latin, Swedish, French, German and English. Most entries are in Celsius' own forward-sloping hand, with occasional additions by some of the hosts and peers he met along the way. It is a time capsule that captures the maturing of a modest but confident young man as he encountered some of the continent's most revered scientific dynasties and locations. In the leaves of this battered book, Celsius blossoms.

Celsius, Biurman and Meldercreutz set off from Uppsala in

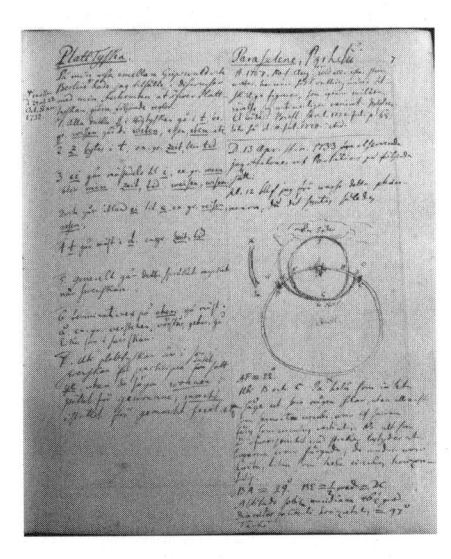

Notes and sketches from Celsius' Grand Tour journal, 1732–36. The author switched between multiple languages in recording his travels.

summer 1732. Their first objective was to catch a mail boat to Germany, but they immediately ran into problems. Having reached the southern coast of Sweden, they discovered that the vessel they hoped to sail on had been ordered to wait for the brother of Prussia's 'Soldier King', Frederick William, to return from his own travels. They were stalled for five weeks, impatient and irritated at the delay, and already eating into their modest budget.

When the tardy Prince eventually arrived, the ship headed out into the Baltic and soon met a squall, which blew them off course, westward towards the shoreline of Denmark. The captain wisely took refuge at anchor for two days, before the winds subsided enough to continue to their destination, Greifswald.

The schisms and boundary changes wrought by the Thirty Years War and Great Northern War had seen this part of Pomerania under shifting jurisdiction for decades. When Celsius, Biurman and Meldercreutz landed at Greifswald, they were technically still within Swedish administration, but culturally and linguistically everything around them was German. Inside the tightly walled city lay its university – established in 1456, over two decades before that in Uppsala, and where Celsius' maternal grandfather Anders Spole had been a student. In the first of his travelling letters

Greifswald – first landfall on Celsius' Grand Tour. Mid-eighteenth-century engraving by Martin Engelbrecht (1684–1756).

to his librarian colleague Benzelius, Celsius explained how (with echoes of both Spole's quick learning on the road and his other grandfather Magnus' mastery of ancient runes) he had managed to acquaint himself with the local language and get by in conversation. With intellectual insouciance, he wrote: 'I constructed a few rules and then I could easily exchange Swedish for Low German.'

After so long stuck at their departure point and the frightening sea crossing, Greifswald was welcoming and comfortable, with the added interest of a mini-pilgrimage to family roots. But Celsius was anxious to move on to their first major destination: the observatory in Berlin. They hired a carriage and headed south across marshy lowlands, for a four-day journey to the Prussian capital.

As they reached the city on 28 October and trundled along the broad Unter den Linden avenue leading to the observatory, the heart-shaped leaves on the lime trees had faded to yellow and begun falling to the ground. The four-storey observatory tower was topped by an iron globe and spiky weathervane, and stood incongruously above the Royal Stables. The interior was filled with the pungent but sweet odour of 200 animals and their stalls below. It was a curious mix of function and location, which mathematician and philosopher Gottfried Wilhelm Leibniz remarked was devoted to both *'musis et mulis'* (muses and mules). The observatory had been completed in 1711 at the behest of Crown Princess Sophia Dorothea, a graceful Hanoverian with a thirst for art, science, literature and fashion.

Celsius soon became accustomed to the oddly pleasant aroma and was impressed both by the observatory's spacious conference room below the tower and its 37-year-old director, Christfried Kirch. Kirch had grown up with astronomy – helping his father Gottfried (the observatory's first director) from an early age, before taking over his responsibilities in 1716. The

An 1824 watercolour of Berlin's royal stables and observatory by Leopold Ludwig Müller (1767–1838).

institution remained a family affair; Kirch had declined repeated overtures from St Petersburg to become the Royal Astronomer there, and had engaged his two younger sisters, Christine and Margaretha, and his mother Maria-Margarethe to support the observatory's work, albeit as unpaid assistants.

Christine and Margaretha demonstrated their skill in using pendulums to measure gravity and the Earth's rotation, and they showed a rapt Celsius records from their studies of Saturn, Jupiter and Venus, comets and the Northern Lights. Meanwhile, the matriarch Maria-Margarethe revealed that she had been making and recording weather observations since 1697 – long before Celsius was born and he and Burman had begun their measurements in Uppsala. She also told him how she had discovered the comet of 1702, as recorded in her husband's diary:

> Early in the morning (about two o'clock) the sky was clear and starry. Some nights before, I had observed a variable star and my wife (as I slept) wanted to find and see it for herself. In so doing, she found a comet in the sky. At which time she woke me, and I found that it was indeed a comet ... I was surprised that I had not seen it the night before.[1]

Berlin's observatory director, Christfried Kirch (1694–1740), and his mother, Maria-Margarethe Kirch (1670–1720).

Celsius admired the purpose and industry of the Kirch family, but could not hide his disappointment that the instruments inside the observatory's viewing tower were few in number and little better than those he already used at home. He noted:

> Mr. Kirch, who is Astronomus Regius, is a good observer with above average experience of practical astronomy, who observes the foremost phenomena that appear, although little encouragement he has there to. We are nights and days together after he is also a pleasant man not too much German. The observatory here is not very well built. Nor very well furnished with instruments. There is no large quadrant here that is sufficient, but small ones that can be used to correct time.

Celsius was interested to hear about how the family financed their studies. In an echo of his late father Nils' enterprise, they earned money by compiling and selling calendars and almanacs. It was an income stream that originated from an edict by King Frederick III, giving the Royal Prussian Academy of Sciences exclusive rights over this important market. And, as the foremost astronomer of the academy, it fell to Kirch and his relatives to deliver. Their products, reliably calculated and released each year, had everyday applications for timekeeping and navigation, and were also emblematic of the age. Almanacs disseminated information, education and discussion – until then largely the preserve of an educated elite – into the public realm. The Kirch clan represented vital aspects of the Enlightenment personified and in action.

Berlin also gave Celsius the opportunity to add to his travelling library and instrument collection. His diary records:

> Here is as good an instrument maker as can be found anywhere in Europe, named Esling. He wants to make an astronomical quadrant for me for 300 Riksdaler, such as de l'Isle brought with him from Paris to St Petersburg. I have already bought the 3rd Volume of Commentariis Petropolitanis.[2]

☼

In spring 1733, Celsius, Biurman and Meldercreutz moved south. Their next stop was the Saxon city of Leipzig. This was home to another university dating back to 1409, from which they continued onto Nuremberg in Bavaria. Here, Celsius began to compile and analyse more than 300 sightings of the Northern Lights from the previous two decades, searching for commonalities in the descriptions for clues as to the cause of this eerie phenomenon. Two pages of his journal transcribe (in English) the experience of Reverend Timothy Neve, secretary of the Gentlemen's Society in Peterborough, eastern England:

> A little after 5 a clock I observed the Northern Hemisphere to be obscured by a dusky red vapour, in which by degrees appear'd several very small black clouds near ye horizon. I thought this seem'd to be a preparation for those Lights which afterwards were seen. The first Eruption of which was within a quarter of an hour, full East from behind one of the small dark clouds. And from after several others full North. These streams of light were of ye same dusky red light as ye vapour just appear'd and vanish'd instantly. I saw eight or ten of these at once about ye breadth of ye Rainbow of different heights several degrees above ye Horizon, and looked like so many red pillows in ye air, and no sooner did they disappear than others shew'd themselves in different places.

Exotic theories about the origin of such spectacular displays were rife. Could they be the work of angry gods or dragons? Celsius himself wondered if erupting volcanoes might be involved – shooting their superheated material high into the atmosphere to be twirled around the rotating planet like

A page from Maria-Margarethe Kirch's almanac for 1701, the year of Celsius' birth.

Frontispiece to Celsius' 1733 Nuremberg pamphlet about observations of the Northern Lights.

glowing spun sugar. He patiently assembled and reproduced the records, not so much to pin his thinking to a particular cause, but to encourage others to share their own observations and ideas. And in years to come, when Celsius had his own observatory and research assistants, this effort would be repaid.

As a sign of his acceptance into the learned community of Nuremberg, the University collated and printed Celsius' aurora borealis accounts. The publication perfectly summed up his emerging scientific philosophy and long-term thinking. Celsius had studied a global phenomenon and suggested how humans should understand and respond to it. Three centuries later, the tone sounds remarkably similar to that of reviews and reports by modern scientists in their struggle to alert humanity to the existential threat of global warming.

☼

By late summer, Celsius was ready to move on again, conscious of how much he and his companions still needed to see and do. The next stage took them 600 kilometres further south, across the Alps and into Italy. The journal does not record how they journeyed through the high mountain passes. It was common for Grand Tourists to make their way on foot or horseback, with the most *effete* travellers being carried on lightweight sedan chairs by strong locals; their voluminous baggage cortèges trailing after them. Celsius and his companions were not of this class, but it is amusing to think of them watching wealthier travellers being borne like this alongside them, and the looks or greetings that might have been exchanged.

The next stop was Venice. The Swedes admired the city's distinctive light and its unique labyrinth of canals, bridges and palaces, but their real interest lay a short distance westward, in Padua. The university there dated from 1222. It was where, in the early sixteenth century, Nicolaus Copernicus had first dared to propose his heliocentric view of the solar system, with the planets in orbit around the Sun.[3] But even this seminal connection could not detain Celsius long – he wanted to reach the next big stop on the itinerary, La Rossa, the 'red city' of Bologna, and its famed, purpose-built observatory, La Specola.

Bologna earned its epithet from the distinctive colour of its bricks and tiles, which seemed to soak up the bright Romagna sunshine and reflect it back in ever-changing shades at different times of the day. Every significant building was constructed from the same, terracotta-hued material, including the university and its observatory, where Copernicus had developed his radical theories.

Polish-born astronomer and clergyman Nicolaus Copernicus (1473–1543), who revolutionised understanding of the universe by placing the Sun in the centre of the solar system, with the planets in plane orbit around it.

Soon after their arrival in mid-October, Celsius, Biurman and Meldercreutz climbed the 272 winding steps of the Specola Mausiliana tower, which sat astride Bologna's meridian at around 11° east. Towards dusk, their hearts pumping from the ascent, they emerged from the gloomy confines of the staircase onto the terrace of the topmost loggia. A sea of tiled roofs stretched before them; a continuous plain of sun-baked clay, with every slope, point and ridge reflecting its own unique, ruddy shade in the fading light. To the south, the folds of the Apennine foothills receded in hazy purplish waves towards Tuscany. The three friends were a long way from Uppsala.

Bologna University was one of the oldest of all – opened in 1088 and in continuous operation ever since. The papal bull that had granted Uppsala permission to develop its university almost four centuries before had conferred the same rights to create faculties and award degrees. Bologna was also home to another powerhouse scientific family: the Manfredis.

La Specola's director, Eustachio Manfredi (1674–1739), had originally studied law before branching out into mathematics, astronomy and poetry. In 1690, aged just 16, he founded the Accadaemia degli Inquieti (Academy of the Unquiet/Restless), which met at his home to discuss emerging theories of science. This later formed the basis for the Bologna Institute Academy of Sciences.[4] Now in his late 50s, Manfredi still retained the intensity of his youth. He had dark, penetrating eyes, topped by heavy eyebrows, and the thickset but agile body of a much younger man.

Manfredi's two sisters, Maddalena (born in 1673) and Teresa (1679) were also highly educated – initially at a convent and then at home, and by working alongside their brother and his collaborators. Together, they formed a tightknit ensemble, building their own telescopes, sextants and other instruments and publishing everything from mathematical tables[5] to collections of Neapolitan verses and folk tales. Maddalena and Teresa never signed their work, and there are no known likenesses of them. But with their seamless grasp of astronomy, mathematics, experimental physics, literature, history and anatomy, they were more than equal partners, deeply immersed in contemporary European culture.[6]

The Manfredis' workplace stood above the sixteenth-century magnificence of the university's Palazzo Pogi headquarters – its interiors a riot of colourful and exuberant decoration. Inside the tower were four floors, each dedicated to part of the family's studies. The first-floor Meridian Room had telescopes aligned precisely along the city's line of longitude. Above this were the astronomers' living quarters and, on the third floor, was the Globe Room, equipped with mural quadrants and more instruments. At the very top, where Celsius and his companions got their first breathless view of the Bologna cityscape, was the Turret Room. This was devoted to plotting the movements of heliacal stars, so crucial to timekeeping and navigation and as signals of the changing agricultural seasons.

Here, Manfredi explored ideas and phenomena that others would later perfect. His work on parallax, the intensity of starlight, aberration of light and obliquity of the ecliptic (the inclination of Earth's orbit to its equator) all steadily accreted proof of Copernicus' theories. He showed that the Earth and other planets orbited the Sun and suggested that light must have a fantastic but finite velocity. With their homemade devices and family enterprise, the Manfredis and their guests were forming the fundamental building blocks of modern astronomy.

Celsius was enthralled by this tower of knowledge, theory and discovery, its functions layered like a sumptuous cake, with its industrious family and staff the delightful filling. He wrote to his mother Gunilla:

> Now I am here in Bologna, and I am feeling quite well, much better than I often did in Sweden. This place appeals to me quite a lot and there are many things to buy. I live here, as far as food in concerned, as if every day were a wedding day. ... The Italians are very polite people towards foreigners. I wish I could exchange Uppsala for this city; in that case I would never leave this place.

The corbelled parapet and turret of Bologna's observatory rising above Palazzo Pogi, surrounded by symbols of the scientific endeavour within.

Director of the Bologna Observatory, Eustachio Manfredi (1674–1739). Engraving by Francesco Rosaspina (1762–1841).

After his purse-lightening travels through Germany, the economy of Italian life evidently pleased Celsius. He first stayed in the house of an artist, close to the observatory and its cavernous gothic neighbour the Basilica di San Petronio. Inside the church, a disc of light from a tiny hole in the vaulted ceiling traced the Sun's progress each day and through the seasons along the meridian path of a decorated marble strip sunk into the floor. On the winter solstice in December, Celsius watched a pale, elongated ellipse creep past the local noon marker. And every day thereafter, he saw its passage grow steadily brighter and rounder.

He then lodged nearby with a Signora Barbara, enthusiastically telling his mother: 'My room, food twice a day, lights and firewood and bed etc altogether do not cost more than ten ducats per month.' The tone of Celsius' correspondence from this time, and the fact that he stayed in touch with Signora Barbara for several years after his stay, hint at some possible romantic attachment – to the culture of his surroundings, if not his hostess. But whether personal ardour was realised, requited or not, this period at close quarters with his landlady and the refined Manfredi sisters certainly lit a flame of some sort.

While out walking one day near the base of Mount Paderno, an extinct volcano on the city outskirts, Celsius noticed and became transfixed by the mysterious text engraved on a Roman-era tombstone. It told of two lovers, Laelia and her soldier sweetheart Agatho – their relationship and fates an enigmatic riddle. It read:

>Aelia Laelia Crispis
>Not man, nor woman, nor androgyne
>Not girl, nor youth, nor old
>Not chaste, nor unchaste, nor modest
>But all
>Carried off
>Not by hunger, nor by sword, nor by poison But by all
>Lies
>Not in air, nor in earth, nor in water
>But everywhere.
>Lucius Agatho Priscus
>Not her husband, nor her lover, nor her friend
>Not mourning, nor rejoicing, nor weeping
>But all

> Erecting this
> Not mound, nor pyramid, nor tomb
> But all
> Knows and knows not
> To whom he erects it.

Celsius could not get the strange words or their elusive meaning out of his head. When he returned to his lodgings that evening, he sat down to imagine and write an ending to the story. He speculated that after Laelia's wealthy family had forbidden her to marry Agatho, she had eloped and served alongside him, masquerading as a male comrade-in-arms. Exhausted and hungry during a long military campaign, Laelia had then taken her own life after eating poisoned bread, but cleverly made Agatho the heir to her fortune. By the time Agatho returned from the wars, Laelia's parents were also dead, so he inherited their estate. Wealthy but heartbroken, he then had the monument with its cryptic words built there, in memory of his lost love.

This rare, written excursion into fantasy still sits among Celsius' papers in Uppsala University library. He was not the first nor the last to be fascinated by the stone and its story, but his interpretation carries another secret – a final sentence in Celsius' hand, which is no longer legible and will therefore remain unknown.

Over a century before Celsius' visit, Mount Paderno's rock had caught the attention of alchemists. They noticed that, after being heated in coals, it emitted an odd, greenish glow, either from within or by some unexplained reflection of ambient light. Dated between 1602 and 1604, these accounts are the first recorded observation of phosphorescence, due at Paderno to the barium sulphide created when the natural minerals in the rock were exposed to fire. This 'moonstone' and the fanciful supposition it generated in Celsius' mind are an apt metaphor for his winter in Bologna – a stimulating yet soothing time warmed by strong emotions and feminine company.

☼

Spring came, and, as much as he was enjoying the comforts and opportunities of Bologna, Celsius knew that they must press on. The ancient papal states capital of Rome lay 350 kilometres further south, a metropolis hotter, older and grander than any they had yet visited.

When the three tourists arrived, Rome was a city still recovering from the discord caused by the corrupt officials surrounding Pope

Benedict XIII, whose short but inglorious term had ended with his death four years before. The Pope's scheming personal secretary, Cardinal Nicolò Coscia, was at the centre of a heartless, self-serving cabal, the Benventines, who effectively exerted control over Rome's political policy. They had exploited the ageing Pope's infirmity to impose all manner of unfair tolls and burdens on the local population, including taxes on meat, fish and soap. Resentment against these abuses exploded into riots when Benedict died in 1730, aged 81.

The incoming Pope, Clement XII, was a similar age to his predecessor, but sharper and better able to face down the graft around him. He expelled Cardinal Coscia, purged the venal Benventines from his court and reversed many of the greatest injustices and excesses of their simony. With so much antipathy lingering in the air, the atmosphere in and around the Vatican was tense. But Rome remained open to overseas visitors and their different faiths, outlooks and habits.

From an introduction by the Manfredis in Bologna, Celsius connected with 72-year-old Cardinal Gianantonio Davia – an intensely devout man, but one with a keen interest in scientific research. Davia had been educated in canon and civil law, and also served as a soldier in the war between Venice and the Ottoman Empire in 1684. After this, he'd moved to Rome and risen rapidly through the ecclesiastical ranks to become a cardinal. In 1725, Davia donated a high-quality pendulum, a reflecting telescope and a quadrant to the Academy of Sciences in Bologna. He and Celsius came from different generations, cultures and traditions, but they shared the same passion for science and its tools.

Davia invited Celsius to his home to observe a solar eclipse, lent him his instruments and even arranged for him to have a private audience with Pope Clement. Bizarrely, the details of this meeting between the gaunt holy father of the Catholic Church and the brilliant young astronomer from Protestant Sweden are not recorded, in Celsius' journal or anywhere else. But whatever was said and transacted between them must have gone well, because Pope Clement offered Celsius and his assistants the use of a local palace with lofty windows, perfect for prolonged observation of the night sky. Celsius' talent for making international friends and attracting their benevolence was in full flow.

In May 1734 he wrote to his cousin and theology professor colleague, Magnus Beronius, at home in Uppsala:

> I have now been here in Rome for a month, and plan to stay here over the summer. The present Pope is very polite towards foreigners and

likes speaking to them. ... I now use a part of my time to study the Roman antiquities which I originally had never intended to do. ... This Italian climate is fully according to my temperament. If things were different I would like to stay here for several years. ... But carriages and horses here are expensive; every time I use a carriage I have to pay twelve pauli, and without a carriage you cannot manage since this city is enormously large.

Despite the generosity of Cardinal Davia and Pope Clement, Celsius was disdainful of the wider clerical scene that infused every aspect of life in Rome. 'Monks and abbots are useless people, walking about in the streets, flirting, talking news at the coffee houses', he continued in his letter to Beronius.

The two years Celsius spent travelling through Germany and Italy embedded many of the themes that would preoccupy him for the rest of his life. They also gave him a vision of the sort of observatory he hoped to create at home and allowed him to collaborate with some of the countries' foremost scientific figures and families. All this made him eager for more. The personal support and approval of the Vatican were flattering, but no substitute for the kind of advanced research and new ground he wanted to explore.

To see, learn and be inspired still further, Celsius knew they must head north and west, to Paris. He had long anticipated that the French capital would provide the intellectual pinnacle of his tour. What he did not yet know was that their arrival would coincide with the greatest scientific issue of the age also coming to the boil. Nor that he would soon be thrust into the very heart of its controversial conclusion, at the top of the world.

10

IN PARIS

Encounters, Observations and Opportunity (1734–35)

> Luck is where opportunity meets preparation.
> Attrib. Seneca the Younger (*c.* 4 BCE–65 CE)

Celsius left Italy with a heavy heart. He had thrived there – warmed inside and out by the sunny climate, wealth of antiquity and from mixing with some of the nation's religious and scholastic elite. As a fair-skinned Scandinavian accustomed to short summers and long winters, he had been transported to a different and refreshing state by the radiance of Bologna and Rome. But intellectually, he still wanted more – to put himself at the cutting edge of astronomy and royal patronage of science, and to become immersed in the biggest questions of the day. This promise lay north, in the capitals and academic powerhouses of France and England.

This next scene of Celsius' Grand Tour would thrust him onto the world stage, cast him into a strong supporting role in the era's most significant scientific debate and settle his future. It would also connect him with a larger network of talented men and women, both his idols and equals – a firmament of other splendid minds.

Perhaps to ease the pain of departing Italy, Celsius offered one of its rising young stars the chance to go with him to Paris. Francesco Algarotti was a precocious 22-year-old polymath, who had already created a stir as a philosopher, poet, essayist, art collector and opera connoisseur. Clearly not someone lacking in ambition, Algarotti eagerly accepted and joined Celsius, Biurman and Meldercreutz as they left Rome in July 1734, bound first for the bustling port of Genoa. From there, they boarded a ship to skirt along the northern Mediterranean coast to the south of France.

Velvet, fur, erudition and romance – Celsius' young Italian companion, Francesco Algarotti (1712–64), painted by Jean Étienne Liotard (1702–89).

We can only speculate about the reason – or reasons – for inviting Algarotti to join them. He was certainly brilliant, versatile and charismatic, a devotee of Newton's scientific principles and a sparkling conversationalist. As the son of a rich Venetian merchant, Francesco had studied natural sciences and mathematics in Rome and Bologna, and had been experimenting with complicated optics since he was 16. Handsome and urbane, he also exuded a strong bisexual magnetism and seemed to be riding a constant wave of powerful urges.

In years to come, Algarotti became a close confidant and regular correspondent with a galaxy of major figures across Europe, including Voltaire, Pope Benedict XIV and Frederick the Great. In 1740, shortly after he became ruler of Prussia, the last of these wrote a curious poem, apparently in response to Algarotti's assertion that northern Europeans lacked the passion of their southern counterparts. Provocatively entitled '*La Jouissance*' ('The Pleasure' or 'The Orgasm') and addressing Algarotti as 'The Swan of Padua', the poem included the lines (translated from the original French):

This night, vigorous desire in full measure,
Algarotti wallowed in a sea of pleasure.
[…]
Everything that speaks to eyes and touches hearts,
Was found in the fond object that enflamed his parts.
[…]

Divine sensual pleasure! To the world a king!
Mother of their delights, an unstaunchable spring,
Speak through my verses, lend me your voice and tenses
Tell of their fire, acts, the ecstasy of their senses!
Our fortunate lovers, transported high above
Know only themselves in the fury of love:
Kissing, enjoying, feeling, sighing and dying
Reviving, kissing, then back to pleasure flying.
[…][1]

It is uncertain whether Frederick was describing an actual liaison between himself and Algarotti, imagining one, or simply visualising the ecstasy of another. But his words give a strong sense of the young Italian's allure. In summer 1734, as Algarotti journeyed to France with the three Swedes, he must have been an entertaining travelling companion. But within a few years, his attractions and affairs would engulf him in scandal.

☼

An uneventful, two-day sea voyage brought the scientists to Antibes. Today, this is a humming, sun-drenched resort on the Côte d'Azur, but then it was a thickly wooded peninsula defended by the imposing bastions and star-plan walls of Fort Carré. They waited here for a day while a suitable carriage was arranged to take them north to Paris – a journey of almost 1,000 kilometres through the heart of France in the stifling heat of August.

From the cramped, lurching confines of their carriage, the four men gazed out at unfamiliar and ever-changing scenes as they passed over the plateaux and mountains of the Massif Central and past countless lakes, rivers and forests. As the distance covered rose, so did their anticipation – what opportunities and encounters lay ahead in Europe's pulsating centre of science, art and culture? Celsius had heard enticing tales about the possibilities in Paris from Samuel Klingenstierna. Now it was his turn to see and sample the city's atmosphere for himself.

When they finally reached the capital, Celsius received good news: official confirmation that Uppsala University had extended his leave. It meant he could continue the tour while retaining his professor's position and salary. Boosted by this, he took up residence in the fashionable Fauborg St Germain district that now forms part of the Sixth Arrondissement on the left bank of the River Seine. The address is recorded in Celsius' letters as the 'Hôtel de Hambourg' in Rue du Four. 'Hôtel' here refers to one of the city's many private mansions and townhouses, which often boasted ornamental entrances and formal gardens. These had been built in great numbers as the rich occupied this area, and properties like this now lined the street where Celsius arrived to find his lodgings.

On the brink of the French Revolution fifty-five years later, the British caricaturist Thomas Rowlandson drew the scene outside Hôtel de Hambourg. The satirical artwork clearly takes aim at the inequality that was soon to explode into such bloody chaos. But it gives some idea of what must have greeted Celsius as he stepped down from the dust-caked carriage.

With his eye for bawdy grotesques, the artist juxtaposes a refined lady with an elaborate headdress looking out from a first-floor balustrade balcony onto a busy thoroughfare of bawdy activity, crammed with hawkers, porters, fruit and vegetable sellers, horses, carriages and cabs. A bonneted young woman props her right foot upon a basket next to the building, suggestively exposing a gartered knee and capturing the excited attention of a leering gent with a staff and sword.

For Celsius, even after the horizon-expanding sights and sounds of his time in Italy, Paris must have presented a profound culture shock. But any concerns he had were quickly dispelled by meeting his hosts – the mother and sister of the French astronomer, cartographer and mathematician Joseph-Nicolas Delisle. Both Mme and Mlle Delisle were themselves highly knowledgeable and interested in the field of astronomy, but M. Delisle was absent. A decade before, he had been summoned to Russia by Peter the Great to create the school of astronomy in the emperor's capital St Petersburg – a posting that would continue until 1747. Celsius knew enough of Delisle and his work though to recognise someone who shared many of his own fascinations with the natural world and the forces that shaped it.

In Berlin, Celsius had of course already met, admired and enjoyed the company of Christfried Kirch's talented sisters and astronomical assistants, Christine and Margaretha, as well as the Manfredi sisters, Maddalena and Teresa, in Bologna. So he was thrilled to find himself under the same roof as another two lively, engaging and informed women. He wrote to Kirch: 'I begin to believe that it is my destiny that all astronomers I have the honour to know during my travels have their learned sisters.'

Mme and Mlle Delisle were equally impressed by the polite young Swedish professor they welcomed into their home. They arranged for Celsius to see the city's famous monuments, buildings and squares, and helped him to explore the immediate neighbourhood. A few blocks to the east lay the Île de la Cité, an elongated teardrop of land in the middle of

Devant Hôtel d'Hambourg – Thomas Rowlandson's etching from 1789.

The Paris Observatory in the early eighteenth century. The 'Marly Tower' on the right was used to mount long-tubed telescopes and even larger tubeless aerial telescopes.

the River Seine, surmounted by the soaring towers, spire, buttresses and windows of the medieval Nôtre Dame cathedral. And a similar distance south of Rue du Four, Celsius walked through the parterres, lawns and fountains of the Jardin du Luxembourg to catch his first sight of what he had really come here for: the palatial Paris Observatory.

☼

The building stood beyond the far end of the gardens. Octagonal towers flanked the solid slab of its central façade, which was decorated with bas-reliefs of astronomical instruments and lined with tall, arched windows overlooking a raised observation terrace. This was science on a different scale, bristling with *brio* and the authority bestowed upon it by France's previous monarch for over seventy years, the 'Sun King', Louis XIV. What might be possible, Celsius thought, if he was able to make a name for himself here?

The observatory was part of France's l'Académie des Sciences. This had begun in 1666 with just a handful of scholars meeting informally in the King's private library and quickly grown into a scientific dynamo energising every field of human enquiry.

The drive to create an observatory dedicated to astronomy and maritime navigation came principally from the King's forceful Minister of Finance, Jean-Baptiste Colbert. Colbert applied all of his guileful statecraft to secure funding and land from the sovereign to realise his plan. Then on Midsummer's Day in 1667 he invited the academy's foremost figures to assemble on a modest hill just outside the city. And with great pomp, he traced the outline of the new building. The architect, Claude Perrault, who had already worked on the Louvre Palace on the opposite side of the river, carefully aligned the foundations so that the Paris meridian (2° 20' 14") precisely bisected the central hall. Even before the building was finished and filled with the latest optical equipment from Italy and the country's finest academics, it was a swaggering declaration of belief in the importance of science and France's stake in it.

Modern visitors to the observatory can still see and trace the meridian line – marked now by a gleaming brass strip laid in white marble engraved with a map of France, showing its longitudinal path from Dunkerque to Perpignan. Like the arrangement that Celsius admired in Bologna's Basilica di San Petronio, a pinhole window at one end of the Meridian Room is positioned so that the sunlight crosses the metal strip at local noon time each day. And alongside are ornate salons, to which Celsius' introductions soon gained him access to begin attending lectures and debates.

The observatory's founding director was the Italian astronomer and engineer Giovanni Domenico Cassini. He had left his native Tuscany in 1669 and arrived in Paris just in time to oversee final construction of the building. Cassini was not wholly convinced by some aspects of Perrault's designs and insisted on changes to both the towers and the central part of the building to make them more suitable for astronomical observation. But once he was satisfied and settled in France, the job became a lifetime appointment. He adopted the Gallicised forenames Jean-Dominique and served up to his death in 1712 as the King's private astronomer and astrologer, advising on the most opportune timings for shifts in policy or military ventures.

By the time Celsius first entered the Paris Observatory, another Cassini was in charge – Giovanni's son Jacques, who had been born at the observatory in 1677 and elected to l'Académie aged just 17. Shortly after succeeding his father, the new director embarked on an ambitious programme to measure the entire north–south length of the Paris meridian's path through France, and from this to calculate the radius of the Earth. This is the route now memorialised in marble on the observatory's floor.

Cassini drew upon the earlier work of the geodesist and priest Jean Picard (1620–82) – the first person to measure the size of the planet to a reasonable

Founder of the Paris Observatory, Giovanni Domenico (Jean-Dominique) Cassini (1625–1712).

Sir Isaac Newton (1643–1727) experimenting with the direction and spectroscopic qualities of light.

degree of accuracy. Picard had surveyed a latticework of thirteen huge triangles from Paris to Amiens. From his measurements, he had deduced that the Earth's radius at this latitude was (in modern units) 6,329 kilometres, corresponding to 110.5 kilometres for each degree of latitude. Modern methods fix the planet's average radius at 6,371 kilometres, demonstrating that Picard's figures were remarkably accurate.

These findings contributed to another scientific breakthrough across the English Channel. In a dust-moted room overlooking the Great Court at Cambridge University's Trinity College, Isaac Newton drew upon the Frenchman's results to formulate his universal theory of gravitation. For a discipline still closely tied to Church teachings and driven largely by observation, Newton's revolutionary conjectures pushed the boundaries of science and tested the mettle of his peers. He suggested that every mass in the universe attracts every other mass, with an invisible force acting between bodies, inversely proportional to the square of the distance between them. So for the Earth and Sun, the strength of gravity is dictated by their comparative masses and how far they are apart.

The Sun's diameter being 109 times larger than that of Earth, and its mass 330,000 times greater, explained its enormous gravitational effect and how it spun all the planets of the heliosphere into orbit around it. Newton inferred that this mysterious force of attraction must be both instantaneous and infinite – operating everywhere, all of the time, regardless of any object's electrical charge or its chemical make-up.

This staggering (but then unproven) insight was published in 1687 as part of Newton's masterwork *Philosophiæ Naturalis Principia Mathematica*.[2] His theory represented a great unification of how gravity had previously been observed and described on Earth (including the famous story of an apple falling onto the author's head), and the known behaviours of stars and planets.

In 1713, Jacques Cassini revisited Picard's measurements to calculate his own meridian arc. He extended the triangulations all the way down to Perpignan in the foothills of the Pyrenees, and arrived at a planetary radius greater than that indicated by Picard – over 111 kilometres per degree of latitude.[3] When this result was published, a fierce dispute erupted. Newton's theory suggested that the effects of gravity on a spinning planet would make it slightly flatter at the poles, and so shaped more like an orange than a perfect sphere. But Cassini's measurements indicated the opposite – that the Earth was squeezed at its equator and so vertically elongated, like a lemon. Oblate (Newton, orange) or prolate (Cassini, lemon), which was right and how could it be proved? The row was reaching a flashpoint just as Celsius arrived in Paris. It was a quarrel in which he was soon to become personally and permanently enmeshed. He was the right person, in the right place, at the right time.

☼

Celsius' notebook from this period is crammed with meticulous notes, drawings and calculations – a heady blend of material transcribed from existing sources, with pages of his own ideas and commentary. It shows that he explored both sides of the argument about the shape of the Earth, and began to attract attention through his incisive contributions.

Celsius soon caught the eye of the court-appointed mathematician, philosopher and l'Académie director, Pierre Louis Moreau de Maupertuis. He was the 37-year-old son of a wealthy merchant from Brittany, with the effortlessly superior manner, fashion and wit of someone privately tutored and seasoned by a (largely honorific) cavalry commission.

De Maupertuis had entered smoothly into fashionable and influential social circles, and quickly gained the ear and trust of the King. He had also become involved in the controversy over the competing ideas of Newton and Cassini, emerging as a forceful advocate of the Englishman's theories.[4] Contemporaries recognised de Maupertuis' talents and his genuine desire to build bridges between science and the arts. When he visited London in 1728, aged 30, he carried with him an introduction from the French naturalist Bernard de Jussieu, which read: 'The person who will give you this letter is a distinguished academician, his taste for natural history, physics and mathematics will make him recommendable to you. That is why I dare not charge him with a recommendation, which he will acquire easily by his own merit.'[5]

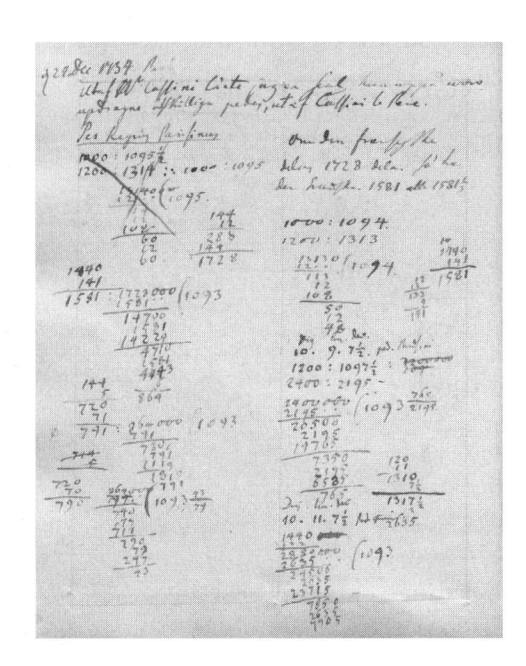

Cross-checking Cassini's calculations – a page from Celsius' Grand Tour notebook dated 24 December 1734.

But de Maupertuis could also be haughty, bumptious and irascible, a man drawn far too readily into arguments and damaging, vituperative correspondence with his rivals. In the early 1730s he fell out with a former close ally, the Swiss mathematician Johan Bernoulli, who chided him for choosing to publish his first major work in England rather than France. And most famously (and disastrously for his reputation) he later attracted the biting satire of Voltaire (1694–1778), who penned a devastating takedown of the mathematician in his pamphlet *Histoire du Docteur Akakia et du Natif de St Malo*.[6] With barely disguised allusion to de Maupertuis, Voltaire wrote: 'We have all too often seen young people, who began by giving great hopes and good works, finally ending up writing nothing but silly things because they want to be skilful courtiers, instead of being skilful writers.' Depicting the mathematician as a despotic buffoon, Voltaire was particularly disparaging about his application of numbers to matters of faith and metaphysics: 'When, in an author a sum of errors is equal to a sum of ridiculousness, nothingness equals his existence ... What a terrible man ... He forges to the left and murders to the right; and he proves God by A plus B divided by Z.'

Determined, debonair and divisive – Pierre Louis Moreau de Maupertuis (1698–1759). He was the director of l'Académie des Sciences, an ardent Newtonian and soon to become Celsius' great friend and advocate.

King Louis XV of France (1710–74) – a young, absolute monarch anxious to continue his predecessor's patronage of science. Portrait by Louis Michel van Loo (1707–71).

King Louis XV of France, *le Bien Aimé* (Louis the Beloved), had succeeded his long-reigning great-grandfather at the age of 5, almost twenty years before. He was a dashing and energetic ruler who was keen to lend his royal benefaction to scientific discovery.

In 1734 he commissioned the advice of another distinguished academy astronomer, Louis Godin, about the best way to finally settle the argument over the shape of the Earth. Godin's father, François, was a parliamentary lawyer. He had hoped that his bright and inquisitive son would follow him into legal practice and so arranged for his preparatory education in the humanities. But Louis took a different turn, developing a fascination first with philosophy and then astronomy, studying under Joseph-Nicolas Delisle at Collège Royal. Aged just 20, he published a seminal set of astronomical tables, which helped him to become a member of l'Académie.

Shortly after his election, Godin made his first presentation about the recent appearance of a meteor, which had terrified many Parisians. This led him to investigate and offer explanations for a variety of other transient phenomena, such as the Northern Lights and, in turn, saw him appointed to edit and index the mammoth, fifteen-volume record of l'Académie's activities, *Histoire et les Mémoires de l'Académie Royale des Science*.[7]

With such a pedigree and standing at the heart of the French scientific establishment, Godin's opinions held considerable sway. When he spoke, others listened. He recommended that expeditions be despatched to the equator and the polar seas to accurately measure a degree of latitude in each location. Comparing the results would then settle the debate over the shape of the Earth. If the planet was a perfect sphere, then the distance of 1 degree would be the same in both places. But a greater distance at one point would reveal where and to what extent the theorised flattening actually occurred.

This was not just about abstract knowledge or empirical evidence – pinpointing the planet's exact form would have far-reaching implications for timekeeping, calendars, navigation, trade and empires. For an imperial power like France with ambitions to spread its influence still further, establishing and owning this proof would confer both glory and advantage. From the King's standpoint, settling this debate was not only advantageous, it was a matter of divine duty. It was the eighteenth-century equivalent of the 1960s space race between the United States and Soviet Union.

King Louis seized enthusiastically on Godin's suggestion of an expedition to the equator, appointing him and two other academy members, Charles Marie La Condamine (a man-of-action soldier, explorer, geographer and mathematician) and Pierre Bouguer (France's leading

Taking a hands-on approach to determining the shape of the Earth: Louis Godin (1704–60), King Louis XV's astronomical advisor and leader of the 1735 equatorial expedition.

hydrographer), to lead a mission to Peru. But mindful perhaps of the potential cost, practical difficulties and more limited territorial benefits, the monarch hesitated about the idea of expeditions to the poles. Since Picard and Cassini had already gathered such rich information from their surveys in France, would the comparison between those and the equatorial measurements not suffice?

De Maupertuis was an astute watcher of the King's moods, and in Louis XV's equivocation he spotted an opportunity. The polar element of proof was essential, he argued, since the greater the distance between measuring points, the larger any disparity in the Earth's curvature would be, and hence the more certain the results. In early 1735, de Maupertuis put a bold proposition directly to the King: with Godin, La Condamine and Bouguer now tasked to go to South America, why not let *him* assemble and lead a similar expedition to the top of the world? If, as de Maupertuis suspected from his own studies and devotion to Newton's theories, the difference in the Earth's radius at the extremes was fairly small, he could offer new methods, new equations and new instruments to detect it. He, de Maupertuis, would assemble a team of the best astronomers, mathematicians and geodesists, and lead them 'to the [North] Pole' to resolve the debate once and for all.

It took months of careful lobbying to persuade the King and, by the time de Maupertuis finally clinched royal approval for his audacious parallel plan, Godin's group had already sailed for Peru. But he was confident he could make up for lost time and put together the necessary personnel and resources. The North Pole was, of course, much closer to Paris than Peru, and, given the distance involved and relatively unknown nature of the destination faced by his counterparts, de Maupertuis was certain he could provide reliable results even sooner. But three big questions confronted him: where to go, how to get there and who to take with him?

☼

De Maupertuis decided to focus on the human resource first. He handpicked a team of other French experts: the 23-year-old mathematician Alexis Claude Clairaut (a former child prodigy with a particular fascination for the properties of curves), astronomer Pierre Charles Le Monnier (even more youthful and full of potential at just 20), and the 35-year-old mathematician and mechanician Charles Étienne Louis Camus. The leader also had the foresight to appoint an official chronicler of the forthcoming adventure, Abbé Réginald Outhier – at 41, a priestly and mature presence among such youth.

But something, or someone, was missing. The chosen team was strong on theorists, but de Maupertuis thought it lacked the necessary expertise in high-precision measurement – especially in the inhospitable terrain and conditions they were sure to face wherever they ventured into the far north. Who could he trust? Perhaps the young Swedish professor he had seen and heard speak so eloquently in recent gatherings at the observatory could assist?

The details of how, where and when de Maupertuis approached Celsius or persuaded him to become part of his expedition are not known. But it is clear that, when the opportunity to support the high-profile, court-funded project arose, Celsius did not hesitate. Yes, he concurred, his experience of fine-grain astronomical and geodesic observations and calculations were exactly what was needed. Moreover, he was already a proven scientific craftsman too, well connected with the finest instrument makers, who could provide the bespoke quadrants, sectors, pendulum clocks and other equipment they would need. Celsius was brought on board, and he quickly helped to answer the other key question of where they should go.

Apart from being from Sweden himself, Celsius could draw upon the valuable experience of his own family and Uppsala colleagues. His grandfather, Anders Spole, had travelled to Lapland in 1695 to study the midnight sun, gravity and magnetism. And, just a few years ago, his university friend and colleague Carl Linnaeus had spent six months in the same region, recording plant species for his seminal work *Flora Lapponica*.[8] Excited and undeterred by his peer's and grandfather's experiences in this primitive and inhospitable land, Celsius suggested to de Maupertuis that the expedition should retrace their steps. The base, he recommended, should be Tornio – the region's largest town at the mouth of the River Torne, which flows down into the Baltic Sea from far beyond the Arctic Circle.

The leader was readily persuaded, signalling the respect de Maupertuis already had for Celsius and his opinions. And he decided to entrust another vital aspect of the expedition to the persuasive Swede: for him to travel to London and secure the very best, custom-designed and custom-built instruments. These had to provide the high level of accuracy they required and be robust enough to stand up to the rigours of the Arctic.

☼

For as long as his money held out, Celsius had always hoped to continue the northbound section of his travels via England. The whole concept of young aristocrats and intellectuals undertaking lengthy foreign tours to expand their minds had originated there, and London was the only city to rival Paris as a centre for astronomical study. But now he could go there with more specific intent, by command of King Louis and with the full backing of the l'Académie director, de Maupertuis.

Inspired by what he had seen in Paris, Celsius also began thinking about the longer term and what he would need after the Arctic expedition, once he returned to his professorship in Uppsala. He asked the University for permission to buy a high-quality quadrant and some other instruments before he left France – a request that, to his dismay, caused 'a hundred difficulties' and prolonged discussion between his colleagues at home. But the money was eventually granted and he was able to place the order for a 3-foot (0.9 metre) radius Langlois quadrant. In a letter to his mother, Gunilla, the only correspondence still preserved from his time in Paris, Celsius wrote: 'I am happy I have been allowed to order the quadrant; it will be ready in the month of August. It was satisfying that I succeeded in that matter; I started to doubt it at the end.' And giving a glimpse of his mood at that moment, he added: 'It is quite difficult to have to leave Paris.'

In just ten months, Celsius' time in the French capital had transformed his prospects and catapulted him into the highest echelons of European science. It was more than he could possibly have hoped for. He was 34 years old, and now – in every sense – a man who was going places.

11

IN LONDON

Appreciation, Trust, Recognition and Craftsmanship (1735–36)

In dreams begin responsibilities.

William Butler Yeats, *Responsibilities*, 1914

The 180-degree panorama from Greenwich Hill remains one of London's finest views. Looking north from the 1676 Royal Observatory, it sweeps away, down across the lush grass and trees of Greenwich Park. The vista hurdles over the classical symmetry of Inigo Jones' Queen's House and the twin towers of the Old Royal Naval College, and then collides with the high-rise office buildings that now occupy London's regenerated docklands. Above, a massive dome of sky forms a natural proscenium arch, framing all the life and drama of the metropolis.

In late July 1735, the ship carrying Celsius from France sailed past this point. He was filled with excitement as he entered London on the final leg of his European study tour. High up to his left stood the Royal Observatory – a place full of fascination and promise. While to his right, where the glass and steel business district now buzzes, were just empty marshes. Ahead, as the River Thames wound into the city, the fluted dome of St Paul's Cathedral – then just a few decades old – towered over everything.

Today, Sir Christopher Wren's masterpiece is barely visible amid the extravagant jostle of modern designer blocks competing for space in the old streetscape. The scene and scale have altered but the grandeur must surely have been the same; Celsius knew that he had arrived somewhere momentous. Appointed to de Maupertuis' Arctic expedition and personally entrusted with obtaining the precision instruments they would need for their adventure, he came to England on the brink of a new and exhilarating stage in his life.

Celsius and Biurman crossed the English Channel together but, for some unknown and apparently urgent reason, Meldercreutz had already returned to Sweden. Just before they left France, Celsius received news that his mother Gunilla was ill. He wrote to Benzelius:

> I am sad to hear dear Mother is not quite well; there is no other remedy for it than patience, since the medics could do very little or nothing about it … Mr Meldercreutz should by now already have talked to dear Mother in Uppsala. It is nearly two months since he left here in a great hurry.

It was over three years since Celsius had left home and he was worried about his mother, who remained the emotional centre-point of his life. As he embarked on this latest and pivotal stage of his travels, he must have yearned for an opportunity to return to Uppsala and the warmth of her love.

Two views of Greenwich Royal Observatory by John Bowles, 1723. The lower engraving shows the River Thames snaking west into the centre of London.

The two remaining travellers had arranged to stay near Fleet Street, a main artery of the capital curving north of the River Thames. It stretched from the old city in the east to the newer royal and government edifices in the heart of Westminster. At this time, a tall stone gateway delineated the two cities. Temple Bar Gate was another Wren design, adorned with statues celebrating the restoration of the English monarchy in 1660 and named after its pre-Great Fire wooden predecessor's function in barring unauthorised trade between the neighbouring authorities. As Celsius and Biurman passed through its central arch they entered a lively locale of taverns, banks, shops, goldsmiths, booksellers and printers, each advertising their services with gaudily painted wooden signs projecting out into the street. The artists of these swinging billboards plundered every aspect of heraldry – real or invented – to furnish the trades beneath with lavish emblems of lions, dragons, falcons and armoured figures.

The new arrivals would have seen the quirky chiming clock mounted on the façade of St Dunstan's Church, on the north side of the street. Every quarter hour, stylised models of two giants struck bells with clubs and then turned their mechanical heads to face onlookers. Below, rumbling coaches splashed pedestrians with filth from the road's central gutter, while criss-crossing sedan chairs, porters and hawkers choked the cobbled passage. It must have seemed a long way from Rome's sun-baked *vias* and the spacious elegance of Paris, but it was vibrant and alive.

Their host here was Dr Cromwell Mortimer: physician, antiquary and secretary of the Royal Society. He proved to be a wise and helpful choice. Mortimer was just a few months younger than Celsius and shared his international credentials, having been educated at Leiden

Etching of London's Temple Bar Gate and Fleet Street around 1740, by F. Wentworth.

University in the Netherlands. The young doctor swiftly set about introducing his guests to an array of renowned people, beginning with his own mentor, benefactor and fellow medic, Sir Hans Sloane.

Seventy-five-year-old Sloane was an engaging and encouraging figure. He too had benefited from overseas travel and study in his youth, with time spent in Paris and Montpelier, and longstanding foreign associateships with l'Académie des Sciences and its equivalents in Prussia, Saxony, Russia and Spain. But there was also darkness and contradiction in the older man's make-up.

In 1687, aged 27, Sloane had travelled to Jamaica to become the personal physician to the British colony's governor. In the oppressive heat and humidity of the Caribbean, he was shocked by the brutality with which the white settlers and plantation managers treated the enslaved West African workforce. At the same time, he was fascinated, recording the violence he saw with a cold, academic detachment.[1] He noted that, for rebellion, slaves were usually punished 'by nailing them down to the ground and then applying the fire by degrees from the feet and hands, burning them gradually up to the head, whereby their pains are extravagant'. He also witnessed frequent castrations, mutilation and vicious thrashings, writing: 'For negligence, slaves are usually whipt. After they are whipt till they are raw, some put on their skins pepper and salt to make them smart; at other times their masters will drip melted wax on their skins, and use very exquisite torments.'

The Enlightenment may just have dawned in Europe, but conditions in imperial outposts remained unreformed and pitiless. For a physician sworn by his Hippocratic oath to do no harm, such scenes must have caused Sloane a different kind of mental agony. But his disgust was counterbalanced by scientific curiosity. Everywhere he went in this tropical island, he came across unknown plants and exotic animals, which he began to collect and document.

In just fifteen months, Sloane gathered more than 1,000 botanical specimens, of which around 800 were species previously unknown to Europeans. He made the first descriptions of the pepper tree and coffee shrub, and experimented with extracting quinine from the bark of the pink-flowered *Chinchona* tree as a treatment for malaria and eye complaints. Sloane's time here even led to him being (somewhat dubiously) credited with the invention of hot chocolate, when he found that adding milk to the raw cacao drink consumed by the indigenous islanders made it more palatable.[2]

When he returned to London, Sloane evidently set aside any reservations about the cruelty and injustices he had seen in Jamaica. He married

The physician, naturalist and collector Sir Hans Sloane (1660–1753) around the time Celsius first met him in London. Portrait from 1736 by Stephen Slaughter.

London's foremost clock and scientific instrument maker, George Graham (1673–1751).

Elizabeth Langley Rose, the wealthy heiress to a sugar plantation owner – a union that enabled him to open a successful medical practice in fashionable Bloomsbury Place and to continue collecting. Celsius might have reflected on the contrast between Sloane's scientific trajectory and his own, brought up in the shadow of his father's frustrations, with only the income from his mother's dining house to support the family.

Sloane's critics mocked him for gathering exotica and curios with apparently little concern for the scientific principles or cultures they represented. One detractor sniped that he peddled 'Nicknackatory', while another dismissed him as 'the foremost toyman of his time'.[3] But Sloane was not discouraged. He amassed an enormous cache of coins, medals, books, manuscripts, hundreds of volumes of dried plants and 'things relating to the customs of ancient times'.[4] He also acquired numerous other collections, absorbing them into his own so that he was soon forced to buy the building next door for more storage. This hoard eventually became the foundation for the British Museum, British Library and Museum of Natural History. In Sloane, Celsius saw not just a man of influence, but someone in single-minded pursuit of his passions – a template he perhaps tucked away to inspire his own later exploits.

With notables like Sloane and Mortimer supporting him, Celsius soon found important doors and exciting prospects opening up. London's most distinguished clock and instrument maker, George Graham, had his premises at the Dial and One Crown, just a short distance along Fleet Street. Graham was a master craftsman and a significant astronomer and geophysicist in his own right. He and Celsius quickly formed a strong relationship – the visitor enthralled by the intricate moving metal orrery models of the solar system, quadrants, sectors and timepieces created by the multitalented artisan. The two men's work became closely intertwined, as they discussed and refined the specifications for the polar expedition's instruments by day and at night observed the planets, stars, solar and lunar eclipses and the aurora borealis together.[5]

Graham's clients included the astronomers at the Greenwich Royal Observatory. On his first visit there, Celsius met the famous and by then elderly Astronomer Royal, Edmond Halley, as well as his successor, James Bradley. Celsius marvelled at the range, scale and quality of the instruments and amenities at Greenwich. In particular, he admired the wood-panelled elegance of the octagonal observation room – another Christopher Wren design – with its built-in wall clocks and decorated stucco ceiling.

Celsius was so grateful for access to these astronomical riches that, at a reception in late October 1735 to mark Halley's seventy-ninth birthday, he lauded the veteran with a profuse Latin address. Celsius proclaimed the astronomer as 'ruling the world', while also taking care to respectfully note that it was the British King's birthday the following day. Despite the gulf in their ages, Halley and Celsius became close associates; the Astronomer Royal even making a number of entries directly into Celsius' journal. In another letter to Benzelius, Celsius wrote: 'I have been to Dr Halley in Greenwich; he is still in full vigour, and is very easy to talk to. We meet every Thursday at a coffee house, where we then eat together.'

London's coffee houses were designed to be islands of civility and equality in the capital's hubbub. They were places where, for the price of a penny, any man (women were not yet welcomed) could mix freely and enjoy lowbrow gossip, virtuoso debate about the natural sciences or anything in between. Customers sat at long communal tables, eavesdropping and interjecting into each other's conversations, while periwigged serving boys raised their jugs head-high to pour dishes of gritty coffee, 'black as hell, strong as death and sweet as love'[6] to fuel the discussions.

An anonymous 1674 woodcut of the *Rules and Orders of the Coffee-House* declared:

Gentry, tradesmen, all are welcome hither,
And may without affront sit down together:
Pre-eminence of place none here should mind,
But take the next fit seat that he can find:
Nor need any, if finer persons come,
Rise up for to assign to them his room.

For Celsius and Halley, the coffee house was a meeting of equals. The two men differed in experience and accomplishment, but were united by mutual regard. Facing each other across the table in a place without alcohol, where cursing, games of chance and talk of sacred things were also banned, the two found the ideal place for their deliberations. The convivial 'penny university' coffee house offered a relaxed alternative to the formal realm of academic life. As Celsius sipped the bitter brew and soaked up wisdom from his venerable friend, his mind no doubt turned to his mother and sister busy at their restaurant back in Uppsala.

The octagonal observation room at the Royal Observatory in Greenwich, designed by Sir Christopher Wren. Etching by Francis Place, 1678.

Bradley became an equally cherished and valuable companion. Despite Celsius' still rather basic English and Bradley's decidedly rough Latin, the Astronomer Royal was able to explain his work on the aberration of light – the apparent motion of celestial objects depending on the relative position velocity of the observer. This is similar to the effect someone experiences in vertically falling rain: when they stand still, the rain drops straight down upon them, but as soon as they start moving, it appears to arrive at an angle. If the person walks briskly into rain, they have to tilt their umbrella forwards, and the faster they go, the more tilt is required.

Bradley showed how the same applies to light rays. The Earth's rotation on its own axis and orbit around the Sun affect how we perceive distant stars and planets. From this, he developed further proof that the speed of light was finite. It was a monumental discovery, which along with his studies of variations in the Earth's axis (caused by the ever-changing gravitational forces of the Sun and Moon) helped to confirm the precise speed of light and the effects of electromagnetism. The rules that Bradley drew up from his experiments meant that these effects could be calculated for any star on any given date, to identify its true, exact position. Modern navigation still relies on his work and it also connects with Albert Einstein's theory of special relativity – the fundamental relationship between speed, mass, time and space.

The second and third British Astronomers Royal: Edmond Halley (*left*, 1656–1742) and James Bradley (*right*, 1692–1762). Both men welcomed and supported Celsius during his time in London.

On 13 September 1735, Bradley took Celsius to meet his uncle and tutor, James Pound. He was the Rector of Wanstead, then a leafy village north-east of London. In the rectory garden, using his host's telescope, Celsius was able to recreate some of Pound's zenith star observations, which had contributed to his nephew's breakthrough work. The Wanstead excursion was just one of many recorded in Celsius' journal to other places and people in and around the capital, most of them with opportunities to sit and observe the night sky alongside fellow astronomers. It was a hectic schedule with apparently little time for relaxation or exploring London's cultural life:

2 October 1735	Observations with Mr Graham in Fleet Street.
4 October 1735	Observations at King Street, Bloomsbury.
23 October 1735	Visit to The Royal Society, Crane Court, Fleet Street.
29 October 1735	Reception for the birthday of Edmond Halley.
11 January 1736	Viewing of eclipses with Dr Mortimer.
15 March 1736	Observations with Mr Graham in Fleet Street.
25 March 1736	Instrument calculations with Mr Graham.
3 April 1736	To Greenwich Observatory with Mr Halley.
11 April 1736	Pendulum experiments to find the time of vibration in a circle with Mr Graham.

In February 1736, Celsius ventured beyond London to visit Clare Hall at the University of Cambridge, where he carried out more observations of the aurora borealis and Jupiter. The courts and manicured gardens leading down to the tranquil River Cam were reminiscent of Uppsala. His diary entries and letters from this period display a twinge of homesickness, but also admiration at the splendour of Cambridge, which he conceded far outshone his own city and university.

On his frequent visits to Greenwich, Celsius noted that here too – as he had seen at the Paris Observatory – ran a meridian, in fact, today's prime meridian, 0° 0' 00". Since 1884 this imaginary pole-to-pole line has been the basis for all international charts, maps and timekeeping. Unlike its polished brass and marble counterpart in France, the London marker is now strangely downbeat and neglected. A modest iron strip heads across the observatory courtyard, snakes down a wall, through a no longer legible eroded sandstone plaque to a narrow walkway, and then disappears off the escarpment edge through some railings and bushes. For 40,000 kilometres the meridian cleaves east from west as it arcs around the globe,

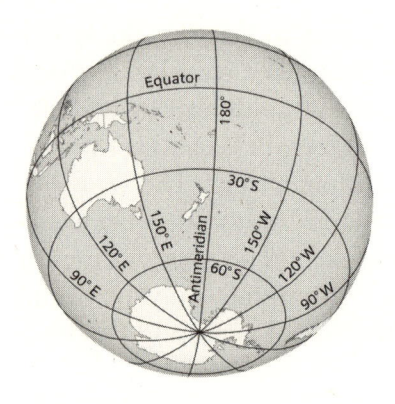

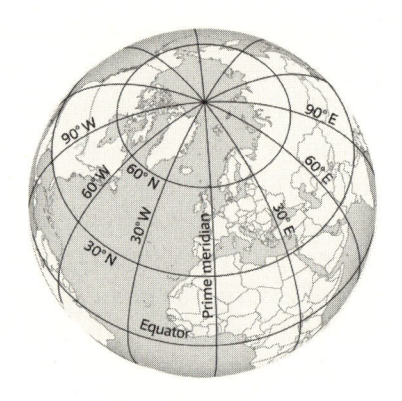

The Earth's northern hemisphere (right) showing the 0° prime meridian passing through Greenwich, and its opposite 180° Pacific antimeridian.

flips at the poles into its opposite 180° 0' 00" antimeridian, bisecting the Pacific Ocean, and then finally catches up with itself back at Greenwich. The International Date Line more or less follows the antimeridian, creating differences in days, not just hours, for long-distance travellers.

Halley, Bradley and Celsius would not have been aware, but in fact this is not the true meridian. Although it indicates the original line on which the first royal observers plotted 0 degrees with a transit circle telescope, modern satellite positioning shows that the actual prime lies about 100 metres to the west. Today, this spot in the park is marked only – and in typically ironic British fashion – by a wonky metal waste bin. As the Royal Observatory's strapline has it, though, Greenwich is indisputably the place 'where east meets west'.

☼

Celsius was so welcomed and fêted by London's intellectual elite that it was perhaps inevitable he would soon be recognised by its premier scientific organisation. An ink-smudged certificate, still preserved in the archive of the Royal Society, records his first meeting there, which marked another important step in Celsius' advance into the uppermost levels of Enlightenment thought. In the firm handwriting of his host and the society's secretary, Dr Mortimer, it reads:

Dominus Andreas Celsius
Astron: Prof: Upsal: & Reg: Soc: Suec: Secret:

[Professor of Astronomy at Uppsala University and Secretary of the Royal Society of Sweden] A Gentleman well skilled in all branches of natural knowledge, as well as Astronomy and Mathematicks, author of a Book intituled CCCXVI Observations de Lumine Boreali &c. of many curious peices printed in the Acta Literaria & Scientarium Suecia & in the Acta Societatis Berolinensis, besides several astronomical tracts in the Suedish language and personally known to most of the learned men in Europe by his late travells thro Germany, Italy & France, and now come to visit England, being desirous to become a member of this Society, We, whose names are under written, thinking him a very fitt and proper person, recommend him as a Candidate.

The ten sponsors below reveal the extent of Celsius' rapid acceptance by London's learned society. Their names join the dots of his busy time in England and demonstrate the growing range of both his interests and associates. Over two neat columns, in addition to Mortimer and Sloane, are the signatures of existing Royal Society fellows Joseph Ayloffe, James Theobald and Martin Folkes (all renowned antiquarians), James Hodgson (mathematician), John Senex (surveyor, globemaker and cartographer), John Machin (astronomer), Richard Graham (barrister and Comptroller of Westminster Bridge) and John Theophilus Desaguiliers (a Huguenot refugee, experimental natural philosopher and chaplain to the Prince of Wales). As references go, it was a formidable list.

However much it was merited or deserved, Celsius' fellowship was not a forgone conclusion. Annotated alongside these names are the dates of ten subsequent occasions when his candidacy was further discussed, before he was formally elected and admitted to the Society on 29 January 1736. Through this process and from these esteemed men and other fellows, Celsius steadily absorbed more knowledge, his self-assurance developing all the time. To augment his rapid recognition in Paris, he had now taken a seat at the top table of British science.

The Royal Society's motto '*Nullius in verba*' ('take nobody's word for it') was an apt expression of the approach to learning that Celsius shared with his new British acquaintances. It was a forceful statement of their determination to withstand the dominance of authority, and to verify things through experiment and an appeal to facts.[7] Since leaving Sweden three years before, Celsius had already become a different man. He was older, bolder and carried a new conviction borne of his foreign encounters and intellectual journey through Europe.

> **Dominus Andreas Celsius**
> Astron: Prof: Upsal: & Reg: Soc: Suec: Secret:
>
> A Gentleman well skilled in all branches of Natural knowledge, as well as Astronomy and Mathematicks, Author of a Book intituled CCCXVI Observationes de Lumine Boreali &c. Of many curious peices printed in the Acta Literaria & Scientiarum Sueciæ & in the Acta Societatis Berolinensis, besides several Astronomical Tracts in the Swedish Language, and personally known to most of the Learned men in Europe by his late Travells thro Germany, Italy & France, and now come to visit England, being desirous to become a member of this Society, Wee, whose Names are under written, thinking him a very fitt and proper person, recommend him as a Candidate
>
> London Oct.ʳ 23. 1735
>
> Hans Sloane   John Senex
> Jos. Ayloffe   J. T. Desaguliers
> James Hodgson   Ja: Theobald
> Crom.ˡ Mortimer   John Machin
> Rich.ᵈ Graham
>
> Elected and Admitted Jan.ʳ 29. 1735.6

Celsius' certificate of election as a Fellow of the British Royal Society, 1735, 036. (Photo by permission of the Royal Society)

While Celsius was building fruitful links with Britain's scientific leaders and institutions, Francesco Algarotti had also made his way to London and was forging more problematic relationships. Soon after crossing the English Channel, the cosmopolitan young Venetian, who had accompanied Celsius from Rome to Paris, became embroiled in a chaotic three-way love affair with an English aristocrat, Lady Mary Wortley Montague (1689–1762), and a political writer and courtier, Baron John Hervey (1696–1743). The two other parties in this unorthodox triangle were almost twice Algarotti's age and both married. In Lady Mary's case, her spouse was, to much embarrassment and unease in high places, the British Ambassador to the Ottoman Empire, Sir Edward Wortley Montague.

Tall, beautiful and engaging, Lady Mary was a free-spirited proto-feminist of fluid and lively sexuality. Twenty years before, she had caused a commotion in society circles by eloping to marry Sir Edward. And long before then, in her childhood diary, she declared that she was 'going to write a history so uncommon'. It was an ambition she did her best to fulfil.

With tastes and insights far ahead of the times, Lady Mary had a clear and critical view of how reigning social attitudes hindered women. She was determined to defy convention by living life to her own rules. In her teens, she taught herself Latin to a high standard, composed witty poems, songs and romances, and was then pursued by a series of eager suitors. Mary was entranced by the traditions and colour she witnessed in Turkey with her husband, often choosing to dress in embroidered gowns and bejewelled headdresses evocative of the Middle East.

When Lady Montague returned to England without Sir Edward to raise their son and daughter, London must have seemed dull and uninteresting by comparison. Algarotti offered her excitement and release, and their liaison continued to smoulder for several years after he returned to Italy in September 1736.

The merest whiff of a connection to this scandal had the potential to undermine Celsius' progress, and cool others' attitudes towards him. It appears that he took care to distance himself from Algarotti and stay focused on the Arctic adventure ahead. On 22 November 1735, de Maupertuis wrote to Celsius from Paris:

> I am ready to execute the sector on the plot, and to take advantage of the good will of Mr Graham. We also need one of these astronomical pendulums, and the machine to measure the pendulum. We have other

pendulums from Mr le Roy and the rest of our Instruments will be made by Langlois. We will leave at the commencement of March, and iron out embarks in Dunkirk, for us to return to Stockholm. You have a few sea brothers to meet you at the Rendezvous in Dunkirk. And you will find charming people charmed to have the honour of your company even if you do not want to see Paris again, and we leave all together for Dunkirk.

Aside from these practical matters, de Maupertuis' letter also ruminated on the difficulties that might await them in Lapland:

I believe you, Sir, that West-Bothnie will be the most convenient place for our operations: however, it is necessary for certain circumstances in the field that we can be sure that one sees the places. If we can find some Lac, or some other glacial sea of 20 or 30 places in the middle of the North, there are a lot of good things to do, but I do not think it is possible to make a mistake in this. We have news of Mr Godin and his troupe, who have arrived at S. Domingue. They are this hour in Peru: and I'm jealous that we are not so proud of our work either.

The letter makes plain de Maupertuis' anxiety about losing ground against Godin's expedition to the equator. Further delays with provisioning and travel arrangements meant that the planned March departure date came and went, with a further two months passing before the polar expedition could finally sail north. Surrounded by the exquisitely engineered ticking clocks in Graham's workshop, Celsius must have been all too aware that time was already their enemy.

In late April 1736, with Graham still hard at work manufacturing the innovative instruments that would be vital to the expedition's success, Celsius set sail from London for Dunkerque. He needed to link up with his scientific colleagues and French sea brothers to complete their mission in the Arctic wilderness. He also longed to be reunited – somehow – with his mother.

PART III

LAND

12

TOWARDS THE POLE

The French King's Arctic Expedition Led by de Maupertuis (1736–37)

No man ever steps in the same river twice, for it is not the same river and he is not the same man.

Attrib. Heraclitus (c. 535–475 BCE)

For someone so preoccupied with accuracy, de Maupertuis' declaration of 'an expedition to the Pole' was uncharacteristically imprecise. Almost two centuries before the first verified human visit to the North Pole, going to the very top of the world was of course beyond his party's reach. But they were determined to get close, as far north as was practically possible and consistent with their task to compare measurements of latitude with Godin and his scientists already at the equator in South America.

On 20 April 1736, de Maupertuis and his compatriots left Paris to meet Celsius at Dunkerque and find a ship there to take them to the Arctic. Anticipating the places, people and practical tests they would encounter on their adventure ahead, de Maupertuis added two further members to the expedition. First, he appointed a *dessinateur* (artist), M. Herbelot, to capture a visual record of their experiences in Lapland. And to help look after the trip's finances, communications and logistical requirements, de Maupertuis persuaded the Minister of the Navy, the influential Comte de Maurepas, to lend him his faithful assistant, M. Sommereux, as secretary.

The latter colleague immediately began to prove his worth – messaging ahead to the Commissary of the Navy at Dunkerque to request that he prepare a suitable vessel and all the necessary supplies for the voyage. The expedition's official scribe, Abbé Outhier, also began his work, documenting the week-long journey north from Paris. With their precious instruments packed in hay to protect them from the constant jolts of rutted and bumpy roads, the party trundled past fortified towns, castles and forests. And their passage through France finished with a short, waterborne stretch 'by very pretty canoe'[1] along the inland canal from Bourbourg to Dunkerque.

As soon as they arrived in the port, among the clamour of the busy harbourside they sought out the vessel that was to take them to the Baltic Sea. Surrounded by the shouts of stevedores and sailors in a dozen different tongues and the sounds and smells of all manner of animals and cargo, de Maupertuis and his team located their ship, *Le Prudent*. It was 'small but strong, and supplied very abundantly with everything that could be necessary for us', observed Outhier.[2]

Le Prudent was a brigantine. She was twin-masted, with the characteristic mix of square-rigged sails on the foremast and triangular sails on the taller aft mast that made these perky little ships so manoeuvrable and dependable for journeys like the one they were about to take. De Maupertuis liked the simple lines and sound condition of what he saw. He also quickly established a good rapport with *Le Prudent*'s captain, François Bernard, and Adam Guen, the pilot who would guide them through the narrow straits and bewildering archipelagos of southern Scandinavia. They were both mature and experienced seamen who inspired confidence among the travellers.

The next day, the scientists decided to relax and divert themselves at Dunkerque's Sunday fair. And while they were there, enjoying the stalls, novelties and entertainments, back at the dockside a lone figure stepped onto the deck of *Le Prudent*. It was Celsius, freshly arrived from London. The group was complete and ready to depart.

An hour before dawn on Wednesday, 2 May 1736, Captain Bernard ordered the mooring ropes to be slipped and the ship carrying the appointed talent and hopes of the French King nosed quietly out into the Channel. Pilot Guen recorded: 'Wind to the south, slight breeze, wet weather.'[3] It was an unremarkable start to a journey of such significance. On board with the scientists, commander and pilot was a crew of four sailors and an apprentice, plus the five servants and cook whom the scientists had brought with them from Paris.

As if to remind the travellers that they had embarked on an assignment of genuine peril, shortly after they left port and began heading towards the

The harbourside at Dunkerque. Print by Victor Adam (1801–67).

A modern interpretation of an eighteenth-century French brigantine, similar to *Le Prudent*, which carried Celsius and his companions to Sweden in early summer 1736. (Image courtesy of Hervé Coutand)

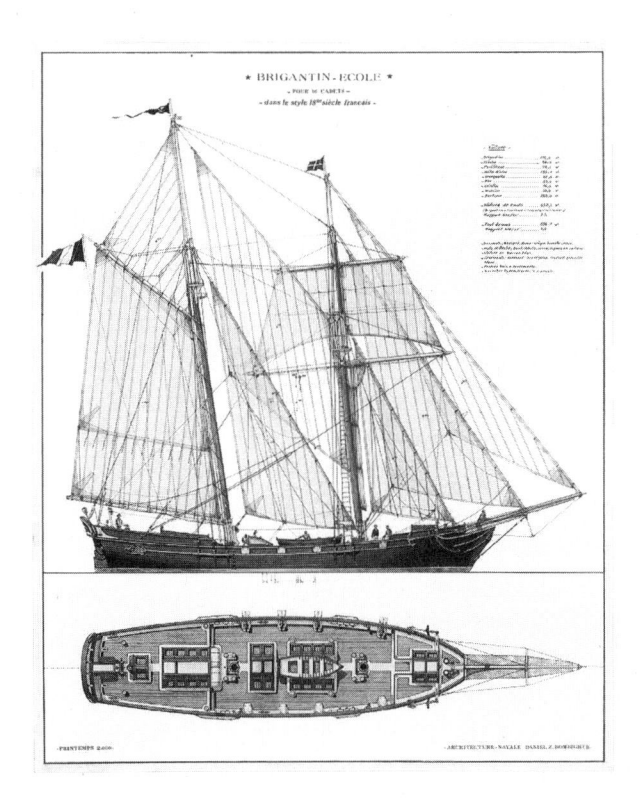

North Sea, a brisk storm blew up. The wind drove them along the coastline in angry, cross-heaving waves that soon felled those unaccustomed to sailing with terrible seasickness. Another feature of the brigantine's design meant that retreating below brought little comfort or relief. Between decks there was little more than a metre of headspace for the passengers in their bunks. The nausea affected everyone except de Maupertuis who, untroubled and keen to demonstrate his *élan*, stood at the bucking prow, gazing fixedly ahead.

For the next nine days, the weather chopped and changed between calmer, drizzly periods and brighter, blustery patches. The scientists and their servants slowly began to gain their sea legs, but the conditions still discomforted them. Outhier noted: 'When the weather was fine, we dined on the bridge: but it was sometimes so leaning that we were ready to roll with our plates, which brought a little disorder to our meals.'[4]

Nature offered some distraction through a series of marine phenomena along their way. Celsius watched on as the sailors took frequent rope soundings of the seabed beneath them to trace their progress and skirt dangerous shoals. Alongside details of every sail setting and plotted course, Pilot Guen methodically noted the material brought up by each sounding: 'Fine white sand – Covered in black shells – Large sand – Like melted cheese' his descriptions read.[5] They also witnessed two impressive displays of the aurora borealis, and on one night the water turned a bright, glistening green with sparks of bioluminescence from creatures beneath them. There were human encounters too: other cargo and fishing craft from England, Belgium, Holland, Denmark, Sweden and Norway occasionally passing by close enough for the crews to exchange shouts of goodwill and reports of sea conditions.

Despite persistent, stomach-hollowing sickness, Celsius was anxious to be well prepared for his duties to come. When the swell subsided enough and the sun emerged from the clouds, he unpacked the small English brass quadrant that he had brought with him and used it to observe and calculate their latitude. He adjusted the viewing scope, mirror and finely graduated scale of the instrument and checked his workings. He was satisfied with the results and reassured of his method, but longed to be ashore.

The passengers' misery increased as their ship rounded the Jutland peninsula and turned south into Kattegat. This shallow, slender passage separating Denmark and Sweden was made even narrower and more hazardous by constantly shifting sand bars, rocky reefs and unpredictable currents. On 11 May the fiercest winds of the whole trip roared down

The route of *Le Prudent* from Dunkerque to Stockholm, in May 1736.

upon them from the north, causing the ship to pitch and roll violently. For Outhier and the others it was a night of unceasing terror:

> The weather was very heavy: we went very quickly and leaning sideways. I stayed all night on the bridge, unable to bear being locked up; I had to hold onto a rope towards the highest edge of the bridge. We were so leaning, that the other side of the bridge was beneath the water.[6]

At daybreak, having judged it finally safe, Captain Bernard brought his ship into the haven of a dark ribbon of water at Helsingør. Beneath the sharp turrets and ramparts of the castle made famous in Shakespeare's *Hamlet*, the travellers gave thanks for their safe arrival and began to recover their composure. For Celsius and Le Monnier it was a decision point. They agreed that after the previous night's horrors they could bear to sail no more. They announced that they would leave the ship and continue to Stockholm overland.

With much relief, the two men crossed over by rowing boat to Helsingborg on the Swedish side and stepped ashore – their balance still upset by the relentless motion of the past ten days. From here they faced a rough and wearying 500-kilometre carriage journey across the

An eighteenth-century view of Kronborg Castle at Helsingør in Denmark, with the hills of western Sweden visible across the narrow sound.

country's interior. But as Celsius and Le Monnier waved off the rest of the team to continue their voyage, plans were afoot for an opulent royal welcome in Stockholm.

King Fredric I and Queen Ulrica Eleonora of Sweden had heard about the expedition heading their way and, influenced perhaps by the involvement of one of their own countrymen in such an illustrious scientific mission, they rolled out the red carpet.[7] On the afternoon of 21 May, a welcoming salute of guns boomed out across the capital's waterline as *Le Prudent* appeared around the final islets dotting the seaway from the archipelago.

The French Ambassador to Sweden, Charles-Louis de Biaudos, Count of Castéja, greeted the scientists as they disembarked. And he immediately set about procuring whatever they might need for the next, crucial stage of their journey northward. The count was a clever diplomat with no qualms about directly intervening in the politics and affairs of the nation to which he was posted. Within forty-eight hours of the two parts of the expedition arriving in Stockholm by land and sea, he had secured an audience for the entire party with the King and Queen.

At this time, the official royal palace at Stadsholmen overlooking the city's inner harbour was still being rebuilt after much of it had been destroyed by a massive fire in 1697. But whatever the monarch's welcome might have lacked in location was compensated for by its warmth, interest and generosity. Outhier wrote with approval how Fredric I spoke in fluent French and 'showed us a lot of kindness'. And he took note that the King was also frank about the hazards and difficulties they could expect to face further north: 'He told M. Maupertuis that we were going to make a terrible trip; that although he had been in bloody battles, he would rather go to the most cruel of all than make the journey we were undertaking.'[8]

More optimistically, Fredric assured his guests that Lapland was 'good hunting country', and he gave de Maupertuis one of his most treasured rifles as a good luck token. The King also decided upon another gift to the expedition. He ordered Sweden's Bureau of Geographical Maps to make available to the Frenchmen all of the latest charts and surveys of their destination – a resource that was soon to become indispensable.

With this royal patronage and diplomatic support secured, de Maupertuis busied himself with the innumerable tasks required before they could continue onward. He skipped between customs offices, banks, stores and suppliers, organising all the documents, money and food they

Enthusiastic supporters of the Arctic expedition: King Fredric I (1676–1751) and Queen Ulrica Eleonora of Sweden (1688–1741).

would need, while sending detailed messages ahead to contacts in Lapland to provide transport, accommodation and practical help once they arrived. Outhier was in awe of his leader's energy and determination to put everything in place, and be prepared for whatever the Arctic might throw at them: 'The length of this trip, the excessive fatigue that had to be endured, the risks there were to be run; nothing was capable of stopping his zeal.'[9]

While equally resolute about their scientific objectives, following their experiences en route to Stockholm the rest of the scientists had no appetite for further sea travel. So it was agreed that the ship would carry on up the Gulf of Bothnia with its cargo and just a skeleton crew of sailors and servants under the supervision of Sommereux, with the rest of the group travelling north overland. Having already crossed the breadth of Sweden, Celsius and Le Monnier were by now accustomed to this option, so they were able to advise on and help secure the heavy, fustian-lined carriages, complete with sleeping bunks, to offer the party as much comfort as possible. Tornio lay over 1,000 kilometres distant, but at least they were now all on dry land.

After a cordial taking of leave from the King and dinner with Ambassador de Biaudos and his wife, the expedition left Stockholm on 5 June. Their first destination was fairly close and, to Celsius at least, familiar: they were on their way to Uppsala. Outhier described the two-day journey as passing through 'a country full of rocks and mounds covered with little firs', and he remarked how 'the paths from Stockholm to Uppsala are beautiful, well-maintained and marked'.[10] This first leg of the journey made a favourable impression on the Frenchmen – they *liked* Sweden, they concurred. And they became even more enthusiastic and appreciative when they entered Celsius' home city, which he had not seen himself for four years.

The governor of Uppsala invited the expeditionary team to dinner – a generous feast at which the travellers enjoyed the ritual of a huge glass goblet filled with white wine flavoured with sugar and orange being passed around the table several times, until it was drained. Celsius then acted as a proud guide to the city's cathedral, university, library, museum and castle gardens.

The visitors from Paris were used to the size and opulence of l'Académie and its observatory, but they were charmed by Uppsala and the well-mannered elegance of its straight streets, wooden houses and graceful, braced bridges across the rushing waters of the River Fyris. They recognised the imprint of this peaceful place in the character of their new Swedish friend, and were able to sample first-hand the environment and atmosphere that had shaped him.

The most touching occasion came the next evening, when Celsius arranged for them all to dine at his mother's restaurant in Svartbäcksgatan. Gunilla's enterprise was by now well established and popular, and, evidently restored to good health herself, she served up a nourishing meal to show her son's esteemed companions how Scandinavian academics fuelled their minds and bodies. In the affectionate glow of Celsius' family, and with their bellies full of interesting food, the Frenchmen saw their Swedish colleague as a whole for perhaps the first time – a treasured son and scientist returning to his roots. Outhier recorded the emotions of their visit and how taken everyone was with this glimpse of their fellow traveller's home life: 'We all went crazy at M. Celsius.'[11]

After so long abroad and away, this flying visit to Uppsala must have been bittersweet for Celsius. He basked in his friends' approval and their appreciation of his ancestral and academic home, but knew he could not stay longer – they had to keep moving north without delay. To ease their progress, de Maupertuis ordered a servant to ride ahead and make sure that enough horses would be available for their carriages at every staging post. In this thinly populated land, some stopping points would only have a single horse and the leader was anxious to avoid wasting time rounding up more animals from fields or woods.

On 9 June, the party left Uppsala and set a punishing pace – travelling all day to arrive at towns late at night, snatching a few hours' sleep and then continuing with fresh horses well before dawn. They passed the burial mounds, ruins and church at Gamla Uppsala and struck out along a straight road Outhier described as 'still beautiful but in country nothing but woods and swamps'.[12]

As Celsius' grandfather Spole had discovered forty years earlier, the biggest obstacles on the journey were the many rivers that streamed down to the Baltic coast from Sweden's central highlands. In some places they found proper ferries, but other rivers had to be overcome by a high-stakes operation of lashing two boats together and balancing the heavy carriages precariously upon them. At each such crossing, the scientists could scarcely bear to watch as the servants and local guides laid down rough planks from the muddy and uneven riverbank to load their precious belongings. Placing the expedition's future upon these narrow boards, they guided the carriage onto the first boat, then gingerly across to the second, the rear wheels clinging onto the creaking, roped platform.

Once cast off, the tension grew even higher as strong winds played havoc, catching the vehicles' high sides and turning the makeshift craft into every direction but the one intended. In this cumbersome fashion, it often took three or four nerve-wracking hours for the whole party to safely pass even modest waterways. Outhier recorded one such crossing over 'a river as large as the Seine in Paris' and 'the roads are very tortuous, and by high mountains whose valleys are almost all lakes or arms of the sea'.[13] And, as the group headed onward, another menace began to trouble them: mosquitoes. These whining menaces forced the travellers to keep the carriage windows shut at night. In the cramped and sweltering interiors, they were barely able to sleep.

Amid these natural hazards and irritants, the scientists found the local population friendly and unstinting in their hospitality. Many times, the Frenchmen felt compelled to pay over the asking price for horses, which seemed absurdly cheap by their standards. And they frequently found it difficult to get the rural Swedes to accept any money at all for the bread, cheese, eggs, butter and fish offered along the way.

The carriages and their occupants pressed on, passing picturesque meadows of barley, rye and millet, and seeing their first grazing herds of reindeer. At a few points, they were also able to spot and briefly contact their ship as it made its way up the coastline in parallel. By the third week of their journey, with 900 kilometres safely completed, the party reached the top of the Gulf of Bothnia and began the last, eastward leg towards Tornio.

At Luleå on 18 June there was a nice surprise. Celsius' Grand Tour assistant, Meldercreutz, had missed his friend's brief return to Uppsala, but set out from there a few days later in pursuit of the group. He was accompanied by a young Swedish lord, Cederström, the son of the Swedish Secretary of State, who was keen to see this remote part of the country for himself. There was no heavy carriage with bunks and windows for these two young men: they rattled all the way in an open, single-horse chaise, with even less protection from the heat and mosquitoes and little time for rest. Outhier made no mention of Meldercreutz's and Cederström's demeanour when they finally caught up with the party in Luleå, but noted that 'their chaise was completely broken'.

For Celsius it was another welcome and heart-warming reunion. He had missed his bright young friend since he had been unexpectedly forced to return home from France the previous year. And he was touched and flattered that he and Cederström had galloped so far and fast to intercept him and wish the expedition well.

The next night brought another novel and encouraging experience. At 11.45 p.m., when they were still on the road, they 'saw the whole sun', bright orange and cresting the horizon. The purists in the party wanted to stop and see if it would still be entirely visible at midnight, but de Maupertuis urged them on and they descended into the shadow of a gulley with Tornio now within their sights.

By an impressive feat of luck or coordination, the crew and contents of *Le Prudent* and the overland group arrived in Tornio within a few hours of each other on 19/20 June. They had completed this long and arduous part of their journey in just two weeks. In another passage of dry understatement, Outhier wrote: 'We were all very tired.'[14] But it was a fatigue tempered by what they had already achieved and the adventure that now lay in store for them.

Unknown to Celsius and the others, half a world away in Peru, two members of Godin's expedition had already perished from disease and accidents and the group had made little progress in measuring latitude. The equatorial party was plagued by misfortune, misjudgement and discord between its key personalities. It would take a decade for the few survivors to eventually return home. For different reasons, what lay ahead in the Arctic for de Maupertuis' contingent would be no less dangerous. It would push the scientists and their helpers to their limits.

Celsius' overland route from Helsingør to Tornio – a hot and gruelling 1,000 kilometres with dozens of perilous river crossings.

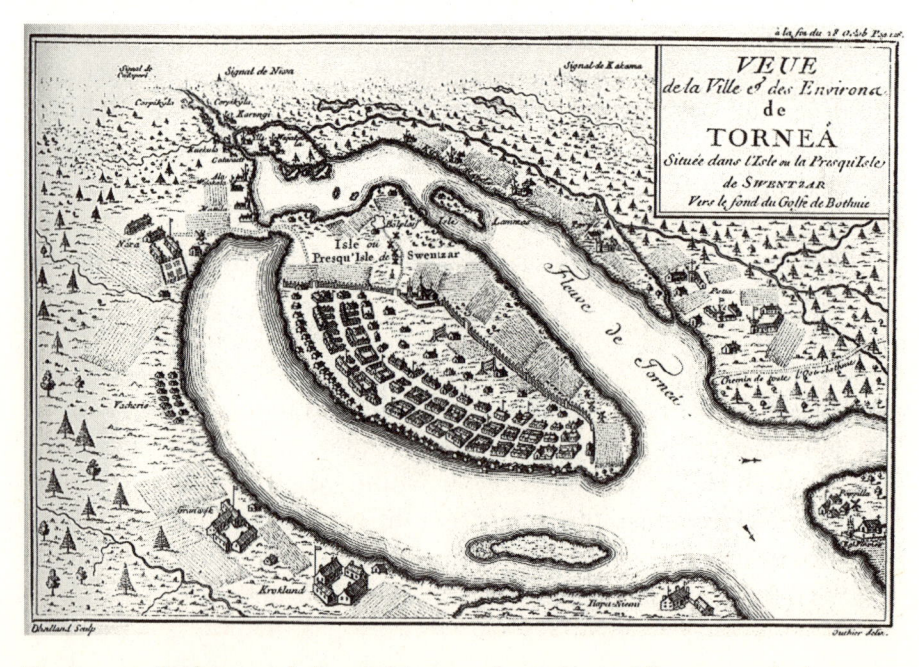

Tornio – until a few years before Celsius came here, the world's most northerly town. This became 'base camp' for the 1736–37 Arctic expedition.

Anders Celsius' paternal grandfather, mathematician and decipherer of the runic alphabet, Magnus Celsius (1621–79). (Courtesy of Gustavianum Art Collection)

Anders Celsius' maternal grandfather – mathematician, soldier and traveller Anders Spole (1630–99). (Image courtesy of Gustavianum Art Collection)

Anders Celsius' father Nils (1658–1724), portrait by Jan Klopper. (Image courtesy of Gustavianum Art Collection)

Celsius' resourceful and stoic mother Gunilla Spole (1672–1756) – the enduring emotional centrepoint of his life. Portrait probably by Johan Scheffel, *c.* 1740. (Image courtesy of Martin Ekman)

Celsius aged 39, after his return from the Arctic. Portrait *c.* 1740 by Johan Henrick Scheffel. (Image courtesy of Martin Ekman)

The ornate compass that Celsius took and used on the Arctic expedition of 1736–37.

The author (left) and Swedish geodesist Martin Ekman at the seal rock on the Iggön peninsula. (Photo courtesy of Pippa Sherval)

The surviving thermometer made and sent by Delisle in 1737, with the new temperature scale annotated by Celsius. What might have happened (or not) if both instruments had broken in transit? (Photo courtesy of Pippa Sherval)

The author (bottom right) unwisely standing on the thawing River Torne ice at the northern end of the expedition's baseline. (Photo courtesy of Mark Hammond)

The author (left) at the top of Mount Kakamavaara with co-founder of the Maupertuis Foundation, Veli-Markku Korteniemi. (Photo courtesy of Mark Hammond)

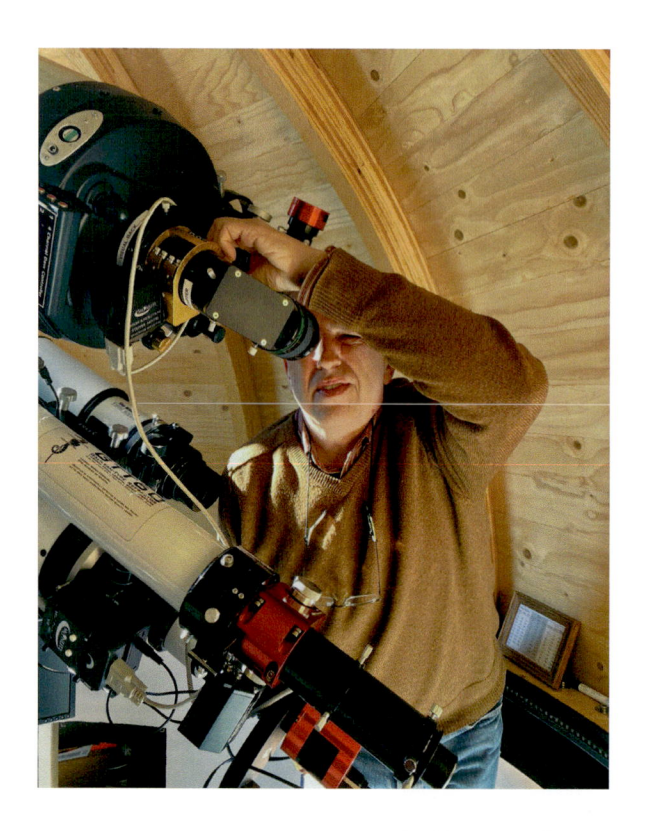

Dutch astronomer and surveyor Nicolás de Hilster in his home observatory near Amsterdam.

The promising young student with a flair for statistics, Pehr Wargentin. Portrait by Carl Fredrich Brander, 1774.

The seal rock (right) at Lövgrund island in the Baltic – with sea levels marked at century intervals since Celsius visited in 1731.

Multicoloured aurora borealis near Tromsø, Norway in 2011.

Modern-day successors to Celsius and Burman maintaining the Uppsala Weather Series.

The warming stripes graphic developed by Professor Ed Hawkins at University of Reading, UK. Each vertical line represents a single year's global average temperature compared to the overall average since 1850.

A burning planet – families watching forest fires in British Columbia, 2003. Climate change is causing the worldwide incidence, scale and destructive power of wildfires to increase at a dramatic rate, turning Earth's natural carbon sinks into the source of enormous carbon emissions. (Photo courtesy of Gary Nylander and the *Kelowna Daily Courier*)

13

ARCTIC SUMMER

The Torne Valley, Instruments and Triangulation

> The greatest achievement of the human spirit is to live up to one's opportunities and make the most of one's resources.
>
> Luc de Clapiers, Marquis de Vauvenargues, *Introduction à la connaissance de l'esprit humain, suivie de réflexions et de maxims*, 1746

The squat stone pyramid nestles among dense mature pines and head-high saplings on the rounded pegmatite hilltop of Mount Kittisvaara in northern Finland. In the cheerful company of Miia Kallioinen and Janne Tolvanen, two employees from the local mayor's office, I came here one early May morning to retrace the route of Celsius, de Maupertuis and their fellow expeditioners.

This modest monument, erected in 1954, marks the final and northernmost measuring point of the group's Arctic exploits. It is where they completed their triangulated measurements and made their concluding observations of the stars to pinpoint the exact length of 1 degree of latitude at this point on the Earth's surface. It lies about 20 kilometres above the Arctic Circle, overlooking the town of Pello, on the border between modern Finland and Sweden.

I wandered a short distance away from my companions, crunching through the calf-deep snow to take in the significance of this spot. I closed my eyes, tuned into the wind whispering in the trees and ran my hands over the pitted face of a huge rock alongside the path – unmoved since it was deposited here by retreating glaciers 12,000 years ago. The eighteenth-century scientists might have passed and perhaps rested against this very boulder as they followed the same path. My eyes alighted on a conspicuous, mossy crevice in the ground, and I knew again that I was standing exactly where Celsius and the other expedition members had walked almost three centuries before.

Turning and picking my way back to Miia and Janne, I saw the pyramid with different eyes – its jumble of irregular but close-fitting blocks

The Maupertuis Pyramid at the summit of Mount Kittisvaaraa, Pello, northern Finland.

and bright green and pink lichens as a symbol of the diverse, tessellating talents of the scientists in whose names and memory it stands. Although barely 3 metres tall, the memorial in this quiet clearing is a highpoint – in every sense – of Enlightenment science. It denotes the climax of a ferocious dispute about fundamental truths, arrived at through an audacious journey into the Arctic wilderness. And it is the precise location of Celsius' individual triumph – his recommendation to bring the expedition to the Torne Valley vindicated and his future secured by the acclaim and lifetime financial support shortly to be conferred upon him by King Louis XV. This is where Celsius arrived and confirmed his place in the first order of scientific history.

The party's route to this point was anything but easy or straightforward. It was a nearly year-long toil full of risks, uncertainty and setbacks. Others would later seize upon some of the group's choices to criticise the mission and its conclusions. But such controversies were yet to come; when the overland and seaborne elements of the expedition arrived in Tornio in June 1736, they were eager to busy themselves and put their combined expertise to the test. It was midsummer, but they knew that, in little more than three months, the ice and winter darkness would return to threaten their wellbeing and inhibit their work.

△△

Having sailed the length of the Baltic Sea, Captain Bernard brought *Le Prudent* into Tornio Harbour and tied up on 19 June. It was a humble port, occupying a finger of land surrounded on three sides by the confluence of river and sea. Three curving streets contained the homes, warehouses and offices needed to support the town's trade, with a thick wall enclosing the whole community.

The next day, while the crew was still unloading equipment and supplies, the scientists also reached Tornio – rattled, tired and dusty after their breakneck journey overland, but elated to see the place that would serve as their campaign headquarters for months ahead. Even in their dirty and travel-weary state, the visitors, with their outlandish dress and manners, must have seemed like exotic arrivals in this remote place.

To assess the length of 1 degree of latitude near to the top of the world with accuracy, the team had to complete three interdependent tasks, each difficult and demanding in its own way. They needed to determine the precise position of specific stars at particular times then measure a baseline on the Earth's surface and survey the angles between fixed points over a distance of 100 kilometres or more. Together, the results would give them the answer they sought: was the planet's curvature here different to that already measured in France or what their counterparts were simultaneously investigating at the equator? If so, how, and by how much did it differ?

The next day's summer solstice provided the perfect opportunity for astronomical observations to fix the base point for measurements. Celsius, Camus and Le Monnier headed straight to the town's Lutheran church in search of a suitable vantage point. They actually found two: the upper, shuttered windows of the freestanding bell tower were ideal for sighting the portable zenith and transit telescope, while the steep, black-shingled spire of the church itself was the obvious landmark to begin the chain of triangles necessary to plot and calculate the full distance of the meridian arc.

Meanwhile, de Maupertuis and the rest of the party were also busily occupied, bobbing across the bay in small boats to explore some of the Västerbotten Islands, which they planned to use as triangulation points. It was not long though before a problem became obvious – these grassy little humps spreading out to sea from the mouth of the river were much flatter than they had expected, and clearly too low to make accurate sightings.

The islands strategy just would not work. But neither could they afford to waste time by waiting for the sea to freeze and then carrying out their measurements directly on the ice of the Gulf of Bothnia. It was an early and crushing blow to the expedition's plans; having spent so much time, effort and a significant slice of King Louis' money getting

Then and now – Tornio church is little changed today from how it looked in 1736.

here, they could hardly just abandon their efforts. They had to devise an alternative scheme – quickly.

Once again Celsius' knowledge came to the fore. From his grandfather Spole's Torne Valley travels in 1695, and his friend Linnaeus' visit here just three years before, Celsius knew that higher mountains and hills lined both banks of the river upstream. And the maps given to them by the King in Stockholm gave some idea of the distances involved. Might these heights provide a solution – assuming of course that they could actually reach them and execute their work in the dense forest?

Ever the activist and leader, de Maupertuis resolved to find out. Mustering some basic provisions, a handful of local soldiers and the expert guidance of the province's governor, Gyllengrip, he left immediately in his boat to reconnoitre possible peaks further up the valley.

Within twenty-four hours, the pathfinders were back in Tornio – their return journey accelerated by the river's strong current. De Maupertuis reported that the highpoints to the north appeared to be perfect for their purpose. He had even climbed one of the mountains – known locally as Aavasaksa – to satisfy himself as to its suitability and to take in the vista of other summits along the valley.

But there were dangers and difficulties too, de Maupertuis warned. Even on the stretch of river so far explored in this brief foray, there were formidable rapids, which had to be avoided by strenuous *portage*. This involved coming ashore, unloading their vessel and carrying everything, boat included, through the thick growth that fringed the shore, to drop back into the water and continue beyond the cataracts. At this time of year there were also insects to contend with – not just a few troublesome mosquitoes and wasps like those they had encountered before, but dense clouds of biting black, warble and botflies, all seemingly intent on vicious airborne attack to bore into any exposed skin. And finally, de Maupertuis explained, there were trees.

While most of the hills were relatively low – 200 metres above the watercourse at most, they were all covered in tightly packed, mature evergreens. At each location, they would have to chop down and remove hundreds of trees in order to create summit markers that would be visible for surveying from adjacent observation points. Even in the clear Lapland air, the quadrants and optics they possessed could only be relied upon to deliver the required degree of visual precision over a few tens of kilometres. This meant locating, reaching and clearing the tops of perhaps a dozen mountains.

It was a forbidding prospect and, having glimpsed the scale of what awaited them, de Maupertuis looked anew at his assembled team. Could they possibly undertake and achieve this task over such an expanse of practically virgin territory within what remained of the short Arctic summer? His fellow scientists were mostly young, energetic and highly motivated. But they were, alas, academics – unused to rowing heavy boats or prolonged physical labour, and with little experience of living and sleeping rough. The accompanying soldiers and servants were, of course, much better suited to the effort and endurance needed, but they were relatively few in number and lacked detailed knowledge of the topography ahead.

Momentarily setting aside his instinctive confidence and can-do nature, de Maupertuis concluded that he should think again before committing to such a risk-laden plan. He returned to his boat and spent the next seven days scouring the coast for places that might support their original design. But his quest proved fruitless. On 27 June, Outhier wrote: 'We had found neither the coast nor the Isles suitable for the work we were here to do.'[1] With a valuable week already wasted, de Maupertuis was out of options. It was the river valley route or nothing.

While their leader was away, the other scientists applied their minds to yet another crucial issue: how and where to establish a baseline from which they could extend all their measurements of linear distances and angles. It needed to be as long and straight as the terrain permitted and as close to sea level as possible. Each point in a survey triangle needed to be visible from the two next to it, and this would be most easily achieved from the hilltops. Similarly, the bigger the baseline (assuming they could accurately gauge its length), the more reliable their calculations should be.

From the cohort of gifted and diverse minds came many suggestions. Le Monnier advanced a plan to slash a swathe through the trees between two points on the most level ground they could find. Once hewn, this line could then work in conjunction with the higher points. It was theoretically feasible and scientifically rational, but completely impractical. Finding and clear-cutting a straight path of sufficient length through dense, mature forest would involve months of unceasing, backbreaking work – time and strength they did not possess.

Again, it was Celsius who provided the answer. Unlike his mid-European counterparts, he was familiar from his upbringing in Sweden with the behaviour and characteristics of frozen rivers. Once the winter ice returned, he explained, it would be thick and stable enough for them to work upon its surface – an accessible and ready-made flat course from which they could obtain precise measurements. With his usual conviction and charm, Celsius reassured his colleagues about the wisdom of the approach, but they were nonetheless anxious about its difficulties. First, they would have to find a stretch of the river long and straight enough to form an adequate baseline, and then they would have to go about their work in almost unbroken darkness and intense cold, surviving on the ice in tents. Even if it could be done quickly, it would test their endurance. Did they have the resolve, energy and resources left for such an undertaking?

The expedition's host and prime logistical assistant in Tornio was the accommodating Mayor Pipping. He introduced the party to Lieutenant Colonel Du Rietz, commander of the West Bothnian Reserves: 'Men of courage, who were not afraid of fatigue.'[2] As well as being brave and tough, the twenty-one soldiers put at the scientists' disposal also spoke Finnish, knew something of the region's territory and, as they had already demonstrated with de Maupertuis' reconnaissance dash upstream, had boats capable of transporting the personnel and all of their valuable equipment and instruments.

In Tornio, a final scientist joined the party – someone with the added advantage of detailed local knowledge. Nineteen-year-old Anders Hellant was a capable student of astronomy and physics from Pello, further up the Torne Valley. Apart from his obvious talent and usefulness in supporting the expedition, Hellant spoke French, Finnish and the native Lap language. He was hired as official interpreter and showed himself to be brave, resilient and inventive – another example of youth at its best.

Almost 300 years later, Tornio's civic authorities extended a similar welcome for my visit. They were keen to show me how the legacy of the 1736–37 expedition remains the subject of great pride in this gateway to Lapland. At Kemi Airport – a tiny, remote terminal a few kilometres east of Tornio, I was unexpectedly met and greeted by Veli-Markku Korteniemi, the genial chairman of the Maupertuis Foundation. Co-founded with his younger brother Tuomo, the Foundation is dedicated to commemorating what took place here and promoting wider understanding of its significance. Veli-Markku was a sprightly 77-year old with sparkling eyes and a neat salt-and-pepper moustache. He explained how his family had very personal reasons to remember the eighteenth-century explorers and their achievements – something that I would see for myself in the coming days.

The next morning, Veli-Markku picked me up from my hotel and drove me the short distance to Tornio's impressive city hall – a square, eight-storey, modern block clad in bay-coloured marble that caught the bright early sun shining from a cloudless blue sky. A replica portrait of de Maupertuis hung next to the lift, which whisked us up to the stylish top-floor council chamber with all-around views out to the coast and up the Torne Valley. On one side, the full-height windows looked over the church spire and bell tower, from where Celsius got his first view of what lay ahead. And from where I stood, several metres higher, the magnitude and ambition of the enterprise also became clear.

Measuring the length of just one out of 180 degrees of the Earth's latitude might not sound much, but when seen from this perspective – with the river snaking away into the distance, its route punctuated by a succession of ever smaller and greyer rounded peaks, the scale of the mission suddenly came into focus. Veli-Markku joined me at the window and pointed out our afternoon objective, off to the right in the middle distance: Mount Kakamavaara. He wanted to show me the sort of unforgiving landscape in which the expedition went about its business.

A few hours later, we parked the car at the side of an unmade mountain road and set off into the snow and fir trees for the summit of this, the third link in the eighteenth-century scientists' chain of measuring points. Despite being almost twenty years older than me, Veli-Markku immediately showed his superior fitness and ease of navigating through what turned into an exhausting and confusing tumble of massive boulders and dark, rocky clefts sheathed in a crust of snow of dangerously unpredictable depth and solidity.

The fitness, he explained, came from decades of ski-orienteering. But his ability to spot and scramble his way up through the tiniest of gaps and ledges looked like something else – something older, unconscious and inherited from past generations living here. So, as I repeatedly plunged, sometimes thigh-deep, into hidden mini-crevasses and clung on desperately to spindly branches for support, Veli-Markku spryly picked his way upward – the modern-day equivalent of Lieutenant Colonel Du Prietz and his troops.

After almost an hour of rough scrambling, we reached the top and turned to look back towards Tornio. Once again, the view highlighted the audacity of what de Maupertuis, Celsius and the others embarked upon. Even with my (admittedly lightweight) modern binoculars, I could only *just* make out the church steeple as a tiny, dark sliver on the southern horizon. And in the opposite direction, another of the measuring points, Mount Niemivaara, was just a fuzzy little bump. The notion of hauling the observation equipment, first through thick forest from the river, then up the way we had come, eaten by insects, and felling dozens of trees to create a sightline and a sturdy wooden marker, filled me with the awe. This was science at the frontiers, hard fought by brave and determined men.

Undaunted, on 6 July 1736 the expedition divided into smaller teams to tackle each of the nine principal summits they had selected, plus a number of subsidiary highpoints. Split between seven boats, the men threw themselves into their task. At the first hill, Mount Nivavaara, they discovered and rapidly perfected the art of stripping bark from pine trunks to reveal bright white wood beneath. When built into vertical cones and tightly fastened at the top, the denuded poles formed ideal markers to stand out on the skyline as measuring points for the quadrants. The surveyors simply needed to set up their instruments inside each of these open teepees, align the sights onto the other visible markers then read off and record the angles to create the triangles.

Despite all the obstacles in their way, the scientists, soldiers and servants forged ahead, the landmarks falling before them like the campaign victories of an unstoppable military advance. By 14 July, they had conquered Mounts Kakamavaara and Huitaperi and, within the next week, work also began on both the highest point, Mount Pullinki, and Aavasaksa where de Maupertuis had scouted a few weeks before. The leader's spirits were so buoyed by their rapid progress that, on the evening of 22 July, he arranged a special dinner at the summit of Aavasaksa, attended by the whole team plus Governor Gyllengrip and the area's most senior pastor, Erik Brunnius. Having made such an energetic and encouraging start to their labours, the expedition was ready to celebrate. In the still-bright night, they sang and finished the last of the wine they had brought with them from France.

In the weeks that followed, the teams kept up the determined

Peeling away bark to reveal bright white wood beneath – perfect material for the summit markers.

pace. By early August they reached and started clearing the northernmost measuring point of Mount Kittisvaara, and also Mount Niemivaara, with its distinctive three-tiered crown of sheer rock walls. Intimidating and exhausting as these peaks were, de Maupertuis was able to appreciate their beauty. He later described Niemivaara in rapturous prose:

> The beautiful lakes that surround this mountain, and the many difficulties we had to overcome in getting thither, gave it the air of an enchanted island in a romance. On one hand you see a grove of trees rise from a plain, smooth and level as the walks of a garden, and on the other you have rocks so perpendicular, so high and so smooth that you would take them for the walls of an unfinished palace rather than for the work of nature. We had been frighted with stories of bears that haunted this place, but saw none. It seemed rather a place of resort for fairies and genii than for bears.[3]

But, at the time, there was no avoiding or disguising the difficulties they faced. De Maupertuis' account of these frenetic weeks talks of them fighting through 'a wilderness of nearly uninhabitable country', 'navigating a river full of cataracts', 'crossing dense forests and profound swamps' and enduring 'the roughest possible walking to scale steep mountains'. And

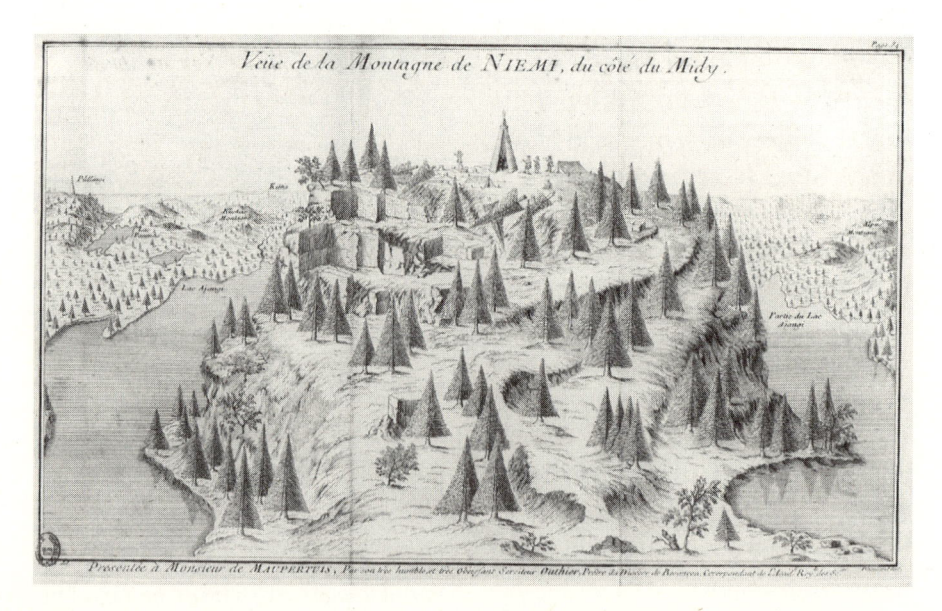

Clearing and preparing Mount Niemivaara, with the white wooden summit marker in place.

perhaps most challenging for all for the members of the group accustomed to the refinements of Paris, he bemoaned them having to 'live in this desert with the worst food and most vicious insects' – 'large, green-headed flies that drew blood wherever they stung us'.[4]

Celsius also communicated some of the Lapland hardships, though in more considered terms. In a letter to his steadfast London friend and supporter Sir Hans Sloane, he wrote:

> I have spent the whole summer with my companions on the mountains of Lapland: where we had a pretty hard time of it. For as there were no houses in our neighbourhood, we were constantly obliged to lie on ye said mountains where our beds and covering were moss and the skins of rene-deer. It was very troublesome to us to mount those rugged hills with our instruments, and to make our way through morasses, marshes and pathless forests without any carriage.
>
> The greater part of our food was salmon, which go up from the Gulph to the cataracts. A prodigious number of gnats has been our greatest inconveniency. Those insects have stung our faces and hands most intolerably. They abound so excessively, that the very Laplanders with their rene-deer are compelled to pass the alps, and reside near the western sea.[5]

On top of the many natural hindrances, there were some setbacks of the expedition's own making. On 18 August, one of the working parties built a tall pyramid of stripped birch trunks to act as the summit marker on Mount Horrilakero – a vital peak further inland from the river, which linked the southern half of the meridian route to a final huge rhombus of observations to the north. Happy with their work, the team descended for the night, but neglected to fully dampen their campfire. Hours later, they watched on helplessly as the embers reignited and consumed both the marker and a substantial portion of the surrounding forest. The next morning, still angry and embarrassed by their carelessness, they climbed the blackened and still-smoking hill to repeat the whole process, this time without a supply of fresh tree trunks readily to hand.

But otherwise, preparation of the all-important measuring points continued with astonishing speed. By the end of August, with all nine main peaks secured, the whole party headed back downstream to the comparative comfort and civilisation of Tornio. Just sixty-six days after the expedition members had first arrived there, this crucial stage was complete. They enjoyed their first change of clothes since setting off up the river in early July and had the chance to rest and reflect on what they had achieved.

Even today, with modern transportation, machinery, equipment and communications, it would be a mammoth undertaking to match what these men did. In the environment and conditions they faced, and with the limited resources available to them, it was nothing short of miraculous. Once they had separated to find and prepare the various mountains, the different groups had no means of contacting each other and no way to raise the alarm if their boats capsized or if one of their number sustained a serious injury. They just had to trust their skills, wits and good luck, hoping that they would all get the job done and return safely. Even before they began the geodesy, mathematics and astronomy they had come here to do, it was a feat of resolve and durability that deserves its own place in history.

A few days after the scientists returned to Tornio, the next stage of their operation was unlocked by the arrival from England of the enormous, purpose-built Graham sector and pendulum clock. These instruments would enable the carefully timed astronomical observations at both ends of the measurement chain to determine the exact latitudes of Tornio and Kittisvaara. And, once combined with the baseline and triangulation results, they would then yield the exact distance of 1 degree as near as possible to the North Pole. Judging if and how this compared to the results from their counterparts in Peru would reveal the exact form of the Earth.

As the servants gingerly prised open the heavy wooden crates, all eyes were on Celsius. It was he who had specified and commissioned the expensive apparatus from the master maker while he was in London, and he was the only person who really knew what to expect inside. When the lids were removed and the protective straw and wadding pulled aside, the hearts of the soldiers responsible for carrying this precious cargo must have sank as they saw what emerged.

The sector was massive, unlike anything they had seen or handled so far. A bulky, braced tripod constructed from solid timber with iron fastenings supported a single vertical wooden limb 3 metres in height. Attached to this was the telescope tube, which could be adjusted with a finely graded scale and micrometer. At the foot of each leg of the tripod was a threaded anchor spike to make the whole contraption exactly level, and in the centre was a freestanding wooden bench for the astronomers to lie back and gaze upward. The local Laps, who later saw the scientists using it, interpreted this strange object as the focus for some kind of idol worship.

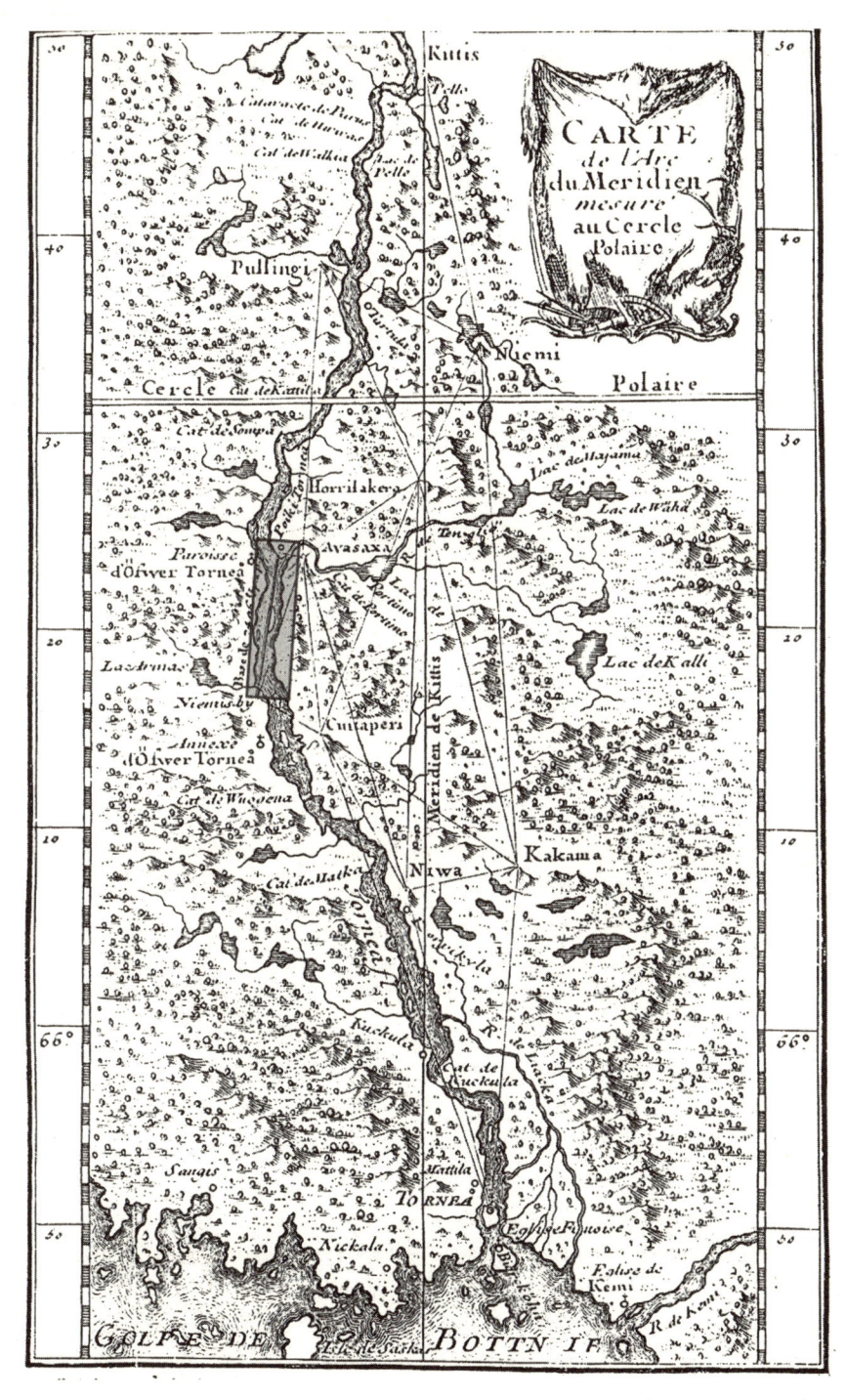

The expedition's completed chain of triangular measurements, from Tornio in the south to Mount Kittisvaara, over 100 kilometres (one degree of latitude) north. The shaded rectangle denotes the frozen river baseline.

For Celsius, the sector's arrival was a quiet but significant triumph. France's most favoured mathematician (and at one remove, therefore, its beloved King) had entrusted him, a young and relatively unproven Swedish professor, with supplying the innovative tools needed for the expedition. And he had delivered. George Graham had personally etched the minuscule graduations on the sector's gleaming brass scale and attached an oval plate with his maker's name confidently engraved in curlicue script: 'Geo:Graham, London'. De Maupertuis was impressed and fulsome in his praise for the object that stood before them, remarking that Graham was 'an ingenious artist', who had 'exerted himself to give it [the sector] all the advantages and all the perfection that could be wished for'.[6]

The short Arctic summer was starting to give way to autumn. There was no time to lose in getting

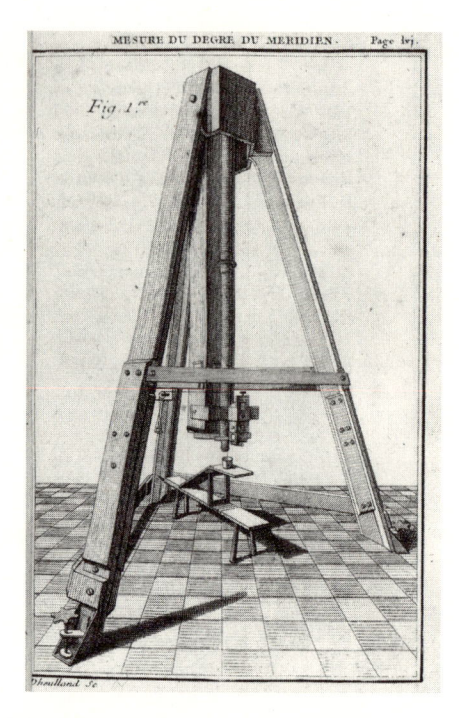

The huge, heavyweight sector built by George Graham to Celsius' specifications to observe stars in the Draco constellation and thereby pinpoint the exact latitudes at both ends of the measurement chain. The viewer lay on the bench in the centre looking directly upwards.

these exciting and important new acquisitions up to Mount Kittisvaara. The party left Tornio on 3 September, spread now across an enlarged fleet of fifteen boats, with three required for the sector alone. Six days later, they arrived in Pello and began constructing two wooden observatories at the top of the mountain – one for the smaller quadrant and a bigger building to house the Leviathan sector. It took three laborious weeks to haul everything to the summit, set it up, check the instruments' calibration and begin making the first star observations.

During this crucial phase, the whole group stayed for the first time at a guest house nearby, at the foot of Mount Kittisvaara in the reedy floodplain alongside the river. The accommodation – a homely contrast to the basic living they had endured so far – was a substantial quadrangle of log lodges with assorted store buildings and byres dotted around the plot. The owners of this welcoming refuge had a familiar name: Korteniemi.

Eight generations before I met Veli-Markku and Tuomo Korteniemi, their ancestors hosted, fed and cared for Celsius and his companions here. So after visiting the summit pyramid I made my way to the site of the guesthouse, accessed now down a narrow lane between scattered houses close to the centre of Pello. Some remnants of the eighteenth-century compound might still be visible if it were not for the terrible destruction wrought by retreating Nazi forces as they were forced from Finland towards the end of the Second World War. In autumn 1944, adopting a literal application of 'scorched earth' tactics, the humiliated Wehrmacht burned and obliterated everything in this part of Lapland: homes, barns, forests, bridges and crops, so that nothing remained for the Finnish population.

The Korteniemi guesthouse was just one casualty of these events, and a sadness seemed to hang over the place as I walked down towards the riverbank, plucking at long, snow-bleached stalks of grass to connect myself to the atmosphere of another moving and meaningful location. A modest, inscribed stone marks where the guesthouse once stood, surrounded by a few decaying shacks. Apart from the now disappeared guesthouse buildings, it was a scene not so different to 1736, with Mount Kittisvaara looming above. A short distance away, across the road and outside Pello's striking black and white modern church, was another stone – a hometown memorial to the expedition's young translator and assistant, Anders Hellant. In this part of the expedition's route, its personalities felt close, familiar and present.

As Celsius settled beneath the Graham sector telescope in the Kittisvaara observatory to take his first views of the Draco stars hundreds of light years away, winter was starting to creep in. On 19 September, patches of ice appeared in the river below them and, two days later, the first flecks of snow fell on the shingled roof. If the scientists did not move fast they could be marooned for months. So, with the preliminary observations complete, everything was disassembled, carted back down the mountain and stowed into the boats for a sprint downstream. In a colder year they might not have made it, but by dodging the thickening ice floes and riding some of the smaller rapids head on, the entire party arrived safely back in Tornio on 28 October.

Within a few days the river was frozen solid, but Celsius and the other astronomers were able to continue their work by building a second observatory on the town's shore to make comparison observations of the same stars. Fixing onto the yellow pinprick of Altais (δ/Delta Draconis), they calculated that they were at 65° 50' 50" north and that the difference between the two ends of their survey chain was 57' 27". It was slightly less than the 1 degree they had planned to measure, but once combined with their triangulations, close enough to give them a definitive answer. Now it was time for the third element of the team's measurements – potentially the most demanding and dangerous of all.

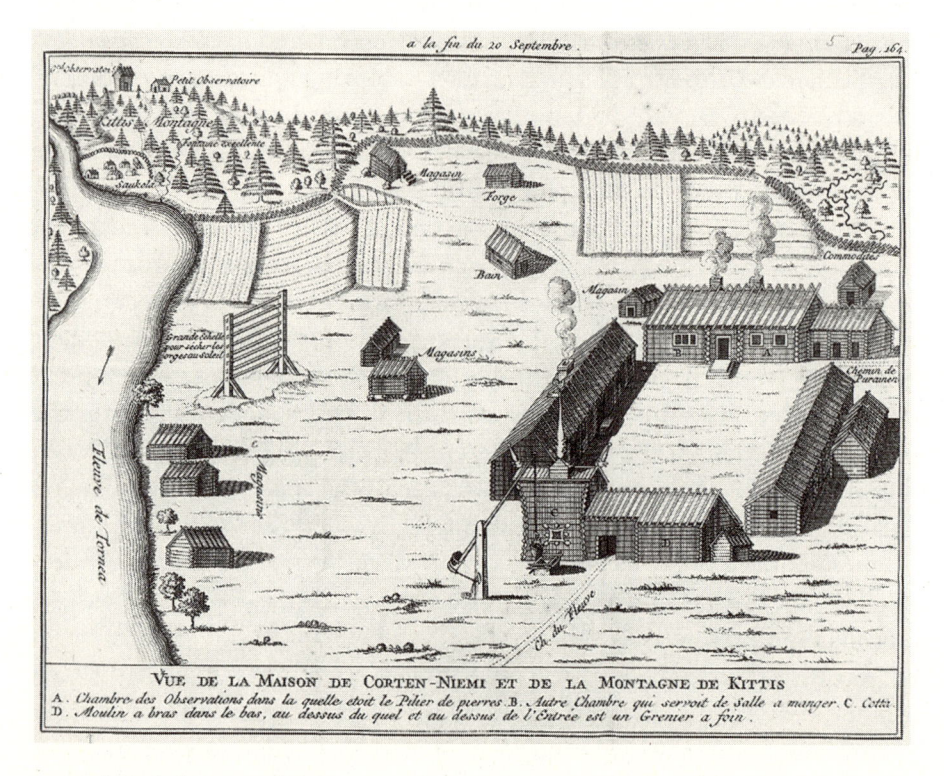

M. Herbelot's drawing of the Korteniemi guesthouse, with the wooden observatories on Mount Kittisvaara top left.

14

ARCTIC WINTER

Ice, Measurement and Calculation

There's a price to pay for speed, and that is danger.
Attrib. Dame Ellen MacArthur

Standing on the frozen river is an odd and unsettling experience. Even close to the bank with at least a metre of ice beneath me, it was hard to dispel the anxiety that, at any moment, a crack might appear or the car-sized slab onto which I had just stepped might upend and plunge me into the black Arctic water. These thoughts crowded into my mind as I sought my balance at the northern end of the expedition's baseline. But my fear – rational, justified or not – was overcome by something else: a compulsion to once again tread where Celsius had been.

Toing and froing between measuring points over the summer months, the expedition pinpointed an ideal stretch of the river as the baseline to validate their calculations and observations. It was a 13.5-kilometre straight line starting at about the midpoint of the valley route, tucked below the sheer cliffs and flat top of Mount Luppiovaara on the western bank. From here it ran almost due north up to a promontory on the opposite bank, close to Aavasaksa and where the smaller River Tengeliönjoki snaked in to add its turbulent waters to the Torne. This is where I stood on an overcast May afternoon, a chilly wind barrelling up the valley and catching at my hair and clothes. I tried to imagine the far colder, near white-out scene that must have greeted the scientists as they ventured onto the ice to begin their measurements in December 1736.

One key difference, of course, was that they were here in almost continuous darkness. The baseline currently lies just south of the Arctic Circle.

And when they started work around the winter solstice, there would have been only a few hours per day of milky half-light and the unpredictable iridescence of the Northern Lights by which to work. The temperatures, even during the short day, would also have been severe – 20 or more degrees below zero, the kind of cold that makes it feel like a hostile Earth is actively trying to kill you, and easily could.

After my stop-off at this end of the baseline, I went to a clean and cosy hotel, with a shower, sauna and soft bed. But Celsius and the others had to stay on the ice, sleeping as best they could huddled together in thin, flapping tents for as long as it took, or as long as they could bear it. I was once again dumbfounded by their immense bravery and fortitude.

Measuring the baseline demanded meticulous planning and preparation. Travelling now on foot and by reindeer-drawn sledges, the party set off from Tornio on 10 December to Övertorneå. Here they divided into two groups to begin measuring from both ends of the line. Camus and Outhier took on the job of making eight long wooden poles, using the cast-iron Langlois *toise* template brought especially from Paris for just this purpose. Since their arrival six months before, it had been kept in a heated room as close as possible to French spring temperatures to achieve the accuracy now required. They gauged and cut each birch length to exactly 5 *toises* (30 feet/9.144 metres), then fashioned and attached simple wooden carrying handles to one side to make them portable. With four poles distributed to each end of the chosen line, the two groups were ready.

In neat, but coincidental, congruence with the expedition's first star observations on 21 June, they began measuring on the winter solstice, 21 December. It was a slow and monotonous process: they had to lay a pole upon the frozen surface, check that it was level and exactly aligned to the distant destination, mark the end of the pole, record the cumulative distance covered, abut another pole to the point reached then repeat the procedure over and over again. Testing in any setting, in the gloom and biting cold of the Arctic winter it was torture. In his 1738 account, de Maupertuis vividly described the experience:

> Judge what it must be to walk in snow two feet deep, with heavy poles in our hands, which we must be continually laying upon the snow and lifting again; in a cold so extreme, that whenever we would take a little brandy, the only thing that could be kept liquid, our tongues and lips froze to the cup and came away bloody.[1]

Alternately sweating then freezing, and struggling to sleep because of the cold despite their exhaustion, the two teams kept this up for ten days straight. They steadily converged on each other, crossing with few niceties as they skirted the long, crooked finger of the Karjosaari islet in midstream, before digging into their final reserves of energy and will to reach the opposite ends of the baseline. For Celsius, it was the ultimate examination of his scientific method and personal tenacity – one that was to bring him both triumph and tragedy.

The frost-mauled measuring teams crawled back into Tornio on the last day of 1736, the bright light, heat and insects of the summer months now a distant memory. With their bodies soon thawed and moods revived, they sat down to compare the results obtained through such extreme labour. What they found exceeded their hopes: the logbooks recorded a difference of just 4 *pouces* (10.8 centimetres) between the two groups' measurements. Over such a long distance and in the harsh winter conditions it was another remarkable achievement. With this exceptional degree of precision obtained on the ground, the men's confidence soared about combining the baseline with their triangulation figures and astronomical

Measuring a degree of latitude with wooden poles by the Northern Lights on the frozen River Torne in December 1736. Etching by J. Ansseau.

The author (right) and Tuomo Korteniemi viewing the expedition's baseline from Mount Luppiovaara on the Swedish side of the River Torne. (Photo courtesy of Mark Hammond)

observations. A definitive answer to the figure of the Earth was within their grasp.

⩓

After six months of non-stop exertion and hardship, 1737 dawned with the expedition members comfortably billeted in various properties around Tornio. The accommodation included one house belonging to the Planström family, two of whose young-adult daughters soon became the centre of a winter scene in which business mixed with pleasure. Few, then or now, would begrudge the scientists and their loyal servants the opportunity to rest, relax and enjoy themselves a little. But in the long winter nights, alcohol and the opposite sex both became increasingly frequent and close companions. There are no first-hand accounts of these events, but it is nice to imagine the scientists and their hosts singing and dancing and literally letting their hair down. It was a short but hedonistic episode, which de Maupertuis' rivals and other critics of his findings would later use against him.

One especially scathing, anonymous retort to the expedition's achievements (published after they returned to France) attacked the team's frivolous 'way of life' in the Arctic, claiming:

> The companions of de Maupertuis, following the example of their leader, each took mistresses; and soon everyone would have no other star to observe than his Christine. There remained of Lapponia only the slight concern for observations and calculations. Every day there were only assemblies, only balls, only colin-maillards.[2]

Whatever went on, the scientists by no means neglected their professional duties. While the thick Baltic ice would prevent their return to France for months to come, there was a mountain of mathematical calculations, cross-checking, pendulum experiments and further observations to complete.

De Maupertuis was puzzled when he examined the initial results derived from collating the three sources of information. Based on what he judged to be the most reliable measurements, the length of 1 degree along the meridian arc was 54,942.57 *toises* (107.08 kilometres). If this was correct, then the extent of the Earth's flattening at the poles would be considerably more than either he expected or Newton's theory had predicted. So, as the days lengthened, the leader ordered further astronomical observations at both ends of the survey chain.

At the Tornio observatory on 17–19 March and back on top of Mount Kittisvaara on 4–6 April, Celsius lay again beneath the giant Graham sector – fixing his attention this time on a different star in the Draco constellation, α (Alpha) Draconis. From the spectral glimmer of this far distant sun he calculated a slightly larger difference between the two points of 57' 30.5", compared to the earlier measurement of 57' 27". Celsius was so confident of this new figure that he described it as having 'a geometrical certainty'. Averaging the two sets of observations put the length of 1 degree on the Lapland meridian arc as 57,437 *toises* – almost 1.5 kilometres longer than that measured along the Paris to Amiens line by Cassini and Picard nearly forty years earlier.

This, it seemed, was conclusive: it meant that the Earth was definitely *aplatie*, compressed at its poles – an oblate ellipsoid, shaped more like an orange than a lemon. Whatever the Peru expedition might already have discovered or would eventually report, the Arctic scientists were sure of their findings. And, once the results had been formally communicated to l'Académie back in Paris, Celsius was able to tell his Swedish peers at the Royal Society of Sciences – in another typically economic turn of phrase – that: 'The Earth's figure will be Newton's opinion.'[3]

Now it was a waiting game. The expedition began packing their equipment and preparing to head south, but they would have to stay in Tornio until the ice cleared enough for their ship to slip out into the Gulf of Bothnia. While they waited, Celsius persuaded de Maupertuis to join him on an excursion beyond Pello, to investigate a rune stone that he had heard about while working up at the northern vertex of their adventure. Uppsala's great eccentric, Olof Rudbeck the Elder, had referred to this 'Stone of Käymäjärvi' in his *Atlantica* writings. From both these sources, Celsius understood that the rock's markings were unusual and quite different to those his grandfather Magnus had deciphered decades before. He was anxious to see this stone for himself – and perhaps discover an entirely new dimension of the runic language and indigenous Sámi culture.

They set off into the forest, more adept by now at steering the odd and unstable one-person Lapland sleigh-cum-boat *pulkas*, and better at managing the volatile moods of the reindeer that pulled them. De Maupertuis recorded the search in a thirty-page account of what became a frustrating and inconclusive excursion, though a memorable one with unexpected benefits.[4]

Well before the Lapland expedition, de Maupertuis had written about what he considered 'three chimeras of science' – quests so elusive and futile that time should not be wasted upon them.[5] The three pursuits he dismissed were 'squaring the circle', 'perpetual motion' and 'the philosophers' stone'. The last idea, a mythical material possessing the power to turn base metals like mercury or lead into gold, had persisted since antiquity. It was the foundation for the ancient practice of alchemy and all of its accompanying folklore. Sceptical as he was, de Maupertuis was nevertheless captivated by the tales of this particular rune stone. Could visiting it be a chance for him, the archetypal empirical scientist, to cross an invisible border into a more primal and mystical world of knowledge?

A local *pulka* sledge like the ones Celsius and his colleagues eventually learned to master in Lapland. A French view of a typical Sámi nomadic home appears in the background – smoky, weatherbeaten and apparently unsuitable for shelter or habitation.

The mysterious markings on the Stone of Käymäjärvi in northern Finland – an unknown runic script or the result of natural causes?

He wanted to find and explore this boundary – if it existed – with his trusted friend Celsius.

In the thick forest and deep snow they at first struggled to find the monument, even with the help of local guides. Eventually, de Maupertuis ordered a fire to be lit in the spot where they believed the stone lay and, as the flames did their work, it suddenly appeared – squat and modest in size, but lent an ethereal glow by the dimming embers and snow-reflected light.

The two scientists stooped and scrutinised the freshly revealed slab from every angle. But neither of them was sure if its seemingly regular scratches and streaks were human-made or just random, natural patterns. They certainly bore no relation to anything Celsius had seen or studied before. Still gripped by the appeal of something preternatural, de Maupertuis speculated that, if these scores were some form of human inscription, it must be one of the oldest on Earth – from a distant, primitive time when the region's climate and population were very different. This turned his imagination to future generations finding a marker and inscription about *his* expedition and its feats in this frozen land. He warmed to the idea of what had, until recently, and without their endeavours, been unknowable, becoming at some faraway time once again unknown.[6] The lonely pyramid at Mount Kittisvaara may yet fulfil his wish.

The two men returned to Tornio intrigued but unenlightened. Not all discoveries, they realised, could be irrefutable. But while evidence of a new alphabet and ancient wisdom had eluded them, something else crystallised between Celsius and de Maupertuis on this journey. They felt a profound ease in each other's company and a mutual respect and affection that went far beyond their identities as scientists. They both sensed that this close union would last and that, when they needed it most, it would serve them well.

On 22 May, shortly after the stone-seekers rejoined their colleagues, some welcome news arrived in a letter from France. Louis XV's representative, Comte de Maurepas, wrote that the King had decided to award Celsius a lifetime royal pension of 1,000 livres per year. It was official recognition of the Swede's crucial part in the expedition's work and a gift that would transform the rest of his life. A livre was originally based on the value of a pound of silver, so this was a sizeable allowance, which at a stroke more than doubled Celsius' income. In modern terms, the pension would be worth around a quarter of a million UK pounds sterling annually. If Celsius' Grand Tour of Europe and his role in the world-shaping Arctic expedition had not already secured his status and future, the French King's largesse would certainly help.

Three weeks later, *Le Prudent* departed Tornio for France. As they left the harbour, de Maupertuis theatrically resumed his position at the prow, pointing the way ahead for the ship and its valuable cargo of geodesic proof. Some sources suggest that the vessel also carried two extra passengers: the Misses Planström, who had either decided (or been convinced) that their futures lay in the company of their French visitors and the glamour of Paris.

Still mindful of the awful seas on the outward journey, Celsius, Camus, Le Monnier and Outhier chose to return overland. Their caution proved to be well advised, since, fewer than 200 kilometres out from Tornio, the ship ran aground near Piteå and had to be abandoned. Thankfully, no one was injured, and the crew, passengers, equipment and priceless expedition records were all safely transferred to continue by land. It took four more weeks for the strung-out party to wind its way back down to Stockholm, where King Fredric and Queen Ulrica Eleonora reprised their royal welcome. At Karlberg Palace, they hosted an extravagant garden party to celebrate the expedition's success and safe return.

It was a joyous but poignant moment. The team would now disperse, with the French scientists continuing on towards France, while Celsius returned to Uppsala in a carriage given to him by King Fredric. He took with him another present: one of the expedition's smaller quadrants. It was a meaningful and useful memento of the astounding feats of which he had been part, and a pointer to his future as a notable professor of independent means. The international *assemblage* of talents had performed gloriously, bonding together through the hazards, labour, stress and fulfilment of their year in the Arctic. But now it was time for them to follow separate paths.

Apart from the two nights on their northward journey, Celsius had been away from his family, home city and university responsibilities for five years. He returned wiser, more assured, financially secure and filled with ambition to build upon his experience abroad. But, deep within his body, the effects of so much punishing travel, graft and sub-zero exposure had inflicted a fatal wound. And – in a portent of bitter twenty-first-century disputes about the veracity of global warming – he was about to become embroiled in a very public scientific storm.

15

FIGHTING FOR THE TRUTH

Conflict and Controversy (1737–40)

He who establishes his argument by noise and command, shows that his reason is weak.

Michel de Montaigne, *Essais*, 1595

The scientists returning from the Arctic brought with them a potent payload of evidence. In just twenty-nine words, de Maupertuis summarised how he believed they had resolved the debate over the shape of the Earth: 'The degree of the meridian which cuts the Polar Circle being longer than a degree of the meridian in France, the Earth is a spheroid flattened towards the poles.'[1]

Having shown 1 degree of latitude in the Arctic to be over a kilometre longer than its equivalent in France, the team was confident of having proved Newton's theories. But they were also conscious that their blend of mathematical and astronomical methods and use of novel, English-made instruments would challenge the traditionalists in Paris, especially the Cassini dynasty. It was not just ideological and theoretical principles at stake; powerful personalities, rivalries and animosities were also in play. As Celsius headed to Uppsala to resume his research and teaching, de Maupertuis set about communicating the expedition's results and lauding himself as its hero. What followed would rewrite the rules of applied mathematics and pull the two men's relationship apart at the seams. It would also reveal a different, more strident and much less appealing side to Celsius' character.

From Sweden, de Maupertuis boarded a ship to Amsterdam, from where he would journey onto Paris. Untroubled again by rough weather making the vessel 'the toy of the winds and the waves', he sent a jubilant letter to Bernoulli in Basel:

> I hope that everyone will be content with this work, and that the question about the shape of the Earth will be forever settled. I will not speak of the life that we had to lead to reach this happy goal, of the cold, the discomfort, the fatigue, the dangers; all that is past, and of fifteen individuals that I led not a single one is dead or ill.

On the same voyage, he wrote to Celsius: 'I beg you that our separation never changes anything in the feelings that you flatter me by having for me, just as I feel that mine will never change.' And in another letter to him a month later, de Maupertuis said:

> I work with all the Swedes I can to compensate for the loss I have made of your company. It is certain that I have become Swedish, or even embothnized in such a way and I do not find myself so well with the French as I thought. I sometimes regret the life we led in Torneå. The losses one makes for one's whole life always have something of what they resemble in death; and I sometimes enjoy the thought of perhaps going to Sweden some day. Keep your friendship for me.

Such tender words reveal the extraordinary closeness between the two men. Conceivably, they might indicate some romantic or sexual intimacy having developed while they were far away. Or perhaps the sentiments were just the opening manoeuvres in de Maupertuis, consciously or otherwise, lining up Celsius to help defend the expedition's work and defeat his detractors.

Once back in France, de Maupertuis went first to the royal palace at Versailles, 15 kilometres south-west of Paris, to brief the King and his representatives. After all the hardships and bleakness of Lapland, de Maupertuis had barely had time to adjust to being back in France and now he was confronted by the garish theatre of King Louis' absolutist court. The sounds of the horses' hooves and iron-rimmed wheels echoed from the walls as his carriage passed through the ornate, crown-topped Honour Gate. He was escorted beneath the tiered crystal chandeliers of the Hall of Mirrors and into a gold-decorated apartment to meet again with the fellow architects of his Arctic adventure, Comte de Maurepas and Cardinal Fleury.

De Maupertuis gave a factual but colourful report of what his expedition had accomplished. The two courtiers listened attentively, their

The Honour Gate (bottom centre) and approach to Versailles in 1682. Drawing by Adam Perelle.

postures and expressions becoming increasingly relaxed as their guest went on. And after the meeting, satisfied with what they had heard, they quickly arranged for de Maupertuis to give a formal presentation to l'Académie des Sciences before it began its annual two-month break in September.

△△

The first Académie hearings took place behind closed doors. While many of the old academicians seemed impressed by de Maupertuis' account, Jacques Cassini immediately took issue with the expedition's approach. In particular, he raised a technical point about the much-vaunted Graham sector. Had they, he enquired, physically turned the instrument in both directions at each observation to ensure its alignment and accuracy, as was standard practice for meridian measurements? De Maupertuis explained that the sector's innovative design meant that this had not been necessary. Nevertheless, the interrogation evidently troubled him, and he ordered Le Monnier to set up and test the sector in his observatory at Collège Royal. He and Clairaut also prepared an urgent written response to Cassini's questions for the Academy's members to consider during their vacation.

Reporting back to Celsius on these tetchy first exchanges, de Maupertuis observed: 'Cassini contained his bad humour a bit in the beginning, but he finally ceased to be the master of it, and raised petty objections against us until the meeting yesterday, when he finally shut up.'[2]

Meanwhile, expedition member Clairaut wrote to Cromwell Mortimer at the Royal Society in London: 'M. Cassini, who has a great deal to lose by these operations, wanted to make some objections; but we responded immediately in such a way that he held his tongue about it, and our Academy is finally convinced of the truth of the flattening toward the poles.'[3]

If de Maupertuis and his team hoped that by now the head of the Paris Observatory had blown himself out, they were to be disappointed. In September 1737, the public got to read about the Arctic results in the popular literary gazette, *Mercure de France*. The newspaper reported that the expedition had 'established the Earth is a spheroid flattened at the poles, as Messrs. Huygens, Newton, and several other great geometers had thought, based on theory'.[4]

In November, de Maupertuis was back at l'Académie for a further hearing, this time an open session in front of a packed audience inside the massive, pillared hall at the Louvre Palace. He drew on all his powers of oratory to tell a stirring tale of the party's privations and triumph in the Arctic, supported by detailed accounts of the team's observations, measurements and calculations. *Mercure de France* was on hand again to report:

> The marked interest of the most numerous assembly there has ever been in any academic meeting sufficiently applauds the work of these illustrious voyagers, the finesse and exactitude of their operations, and the clear and elegant manner in which M. de Maupertuis made everyone capable of judging it.[5]

The entomologist René Antoine Ferchaur de Réaumur, one of the academy's elder statesmen, heaped praise upon de Maupertuis' speech. In a mood that brings to mind the modern-day product launches by technology entrepreneurs, he recorded:

> M. de Maupertuis's report lasted more than an hour and a half, and everyone in the audience found it too brief. The gathering of the audience was prodigious; there could be no thought of closing the doors; part of the gallery was filled with those who were not able to enter into the hall. It revealed that everything was done with the most scrupulous exactitude.[6]

But Cassini would not let things go, and both sides continued to shore up their positions while seeking fresh evidence to support their theories. As

L'Académie des Sciences in session at the Louvre Palace, Paris in 1769.

steadily more vitriolic polemics were exchanged between the two sides, de Maupertuis contacted Bradley and Graham in London, asking for a certificate of the sector's accuracy and details of stellar observations from Greenwich to corroborate those they had recorded in the Arctic. And he also appealed to Celsius for help. On 31 January 1738, de Maupertuis wrote to him in Uppsala: 'You will do very well to make known to the public the negligence and faults of M. Cassini in a matter in which he wants to make us the fools. I have no doubt that you have found in the book enough to make it ridiculous.'[7]

De Maupertuis' letter continued with a detailed critique of Cassini's measurements – not just enlisting Celsius' support, but feeding him the exact quantity and calibre of ammunition with which to bombard his rival's claims. The letter concluded: 'You would do very well to publish your letter sooner and treat our adversaries as they deserve.'[8]

In private correspondence with Benzelius, Celsius had already expressed his unfaltering belief in the Arctic expedition's work. 'Nature would be too bizarre if the Earth would have the figure of Cassini', he wrote. And now, coaxed by de Maupertuis into justifying this confidence in public, Celsius needed little encouragement. His 1738 pamphlet *De observationibus pro figura telluris determinanda in Gallia habitis disquisition*[9] began with an excoriating attack on the shortcomings of Cassini's French-made instruments, followed by a comprehensive denouncement of the home-country measurements, theories and traditions. Celsius wrote: 'Cassini's observations, terrestrial and celestial, in the southern part of France, are sufficiently uncertain that it is impossible to deduce the shape of the earth from them.'[10] The tone was harsh and uncharacteristic, as if Celsius was both stung by the criticism of his own expedition's work and desperate to please his leader and friend. It also seemed to display a newfound conceit and arrogance, born perhaps of his financial independence thanks to King Louis' pension gift.

Celsius' long-time friend, university colleague and correspondent, Eric Benzelius (1675–1743).

Delisle, whose family had hosted Celsius in Paris, wrote to him from St Petersburg, objecting to the pamphlet's petulance: 'I find the conclusion that you draw at the end of your dissertation very rude. Could we not hope for a softening of this position?'[11] But Celsius stood firm. He replied to Delisle: '[Cassini] is surely the aggressor, since he attacked our observations in the Academy and afterwards spread a rumour in the gazettes about the uncertainty of our operations.'[12]

Angry that his family's honour had been impugned, Cassini hit back with two passionate, in-person presentations to l'Académie, which he then published in a densely argued pamphlet of his own.[13] If it had been de Maupertuis' plan to draw Cassini out into an open fight, it worked perfectly. Beyond the accusations traded mostly in the technical language of astronomy, the acrimony spread to the Parisian cafés, drawing rooms and press, provoking yet further conflict. For Celsius though, the situation was all too redolent of his late father Nils' bitter, career-wrecking disputes with the Church of Sweden a generation before. He had seen up close the sad, degenerating effect of being in ill favour with the establishment and, maybe now regretting the aggressive salvoes he had already fired off, he had no desire to repeat it.

With the controversy coming to a head, de Maupertuis prepared to publish his own version of the Lapland expedition and its vindication of

Newton's theories. But before the printers had finished their work, two other events created yet more ripples of discontent.

△△

First, there was news at last from Peru, where the equatorial party's pendulum experiments appeared to support the northern hemisphere conclusions about the shape of the Earth. While de Maupertuis' team had battled against the polar conditions to achieve their goals and make their way safely back home, in South America everything that could possibly have gone wrong had done.

Sailing across the Atlantic, through the Caribbean and then into the steaming jungle rivers of the Isthmus of Panama, it had taken Godin, La Condamine, Bouguer and their team over a year to reach Quito (now the capital of Ecuador). This was the northernmost point of their meridian, perched almost 3,000 metres up in the Andean foothills. On the way, they suffered terribly from the oppressive heat and humidity, as they travelled through uncharted mountains and mangrove swamps in constant fear of alligators, scorpions and disease-carrying insects.

Before they even reached the mainland continent, Godin had spent a substantial part of the expedition's budget on buying a diamond for a prostitute he met at a brothel in Saint Domingue (now Haiti). And while

The equatorial expedition poling its way up Panama's Chagres River in a dugout canoe, surrounded by exotic flora and fauna.

their ship was anchored at Manta Bay on the Pacific coast of Ecuador, the group disintegrated completely – falling out over when and where to begin their scientific mission. A fuming Godin sailed off to Peru, leaving Bouguer and La Condamine, plus two slaves and a servant, stranded on the beach with their instruments.

As various accounts of the South American expedition conclude, the Frenchmen were pitifully ill-equipped for their task. They were 'deskbound mathematicians who hadn't the faintest clue what leadership meant'.[14] Their troubles put the relatively smooth conduct of de Maupertuis' expedition into some perspective.

Things deteriorated further when the expedition's surgeon, Jean Seniergues, fortified maybe by too much *aguardiente*, unwisely accompanied a local young woman who had become his lover to a bullfight in the city of Cuenca, at the southern end of the one-degree arc. The woman's father was sitting right next to her and a violent mêlée broke out, culminating in Seniergues being surrounded by an irate mob armed with pikes and lances chanting, 'Kill the French foreigners!'[15] And kill they did – Seniergues died a week later from dozens of stab wounds inflicted during the fight.

Despite all these quarrels and setbacks, the group managed to reconvene, bury their differences and thread a survey of not 1 but 3 degrees of latitude through the unforgiving landscape. Their measurements, since shown to be accurate to within about 45 metres, showed that a degree at the equator was significantly shorter than had already been surveyed in

The mass brawl at a Cuenca bullfight, which cost the life of the French equatorial expedition's surgeon, Jean Seniergues.

The equatorial meridian survey in Peru. The leaders, La Condamine, Godin and Bouguer, fought against vertiginous heights, hostile tropical conditions and each other.

France and – although they did not yet know it – by de Maupertuis' team in the Arctic. As news of the Peru findings filtered back to Europe, the case for Newton's theories gathered strength; everything pointed to the Earth being flattened at its poles, by a factor of around 1 in 300.

The second event to stoke the flames of disagreement was the arrival in Paris in November 1738 of Christina and Elisabeth, the Planström sisters from Tornio. Exactly how they got there is unclear. Reports differ about whether they had continued all the way from Lapland to Stockholm with de Maupertuis once *Le Prudent* had foundered on its return voyage, or if they somehow made their own way to the Swedish capital. But once there, they were evidently determined to keep pursuing the scientists on to France. *Les lappones* (as they became known) quickly showed themselves to be both adroit and resourceful in negotiating the privileged ways of Paris.

In early 1738, de Maupertuis wrote to Celsius, saying that he had left in Stockholm 'two poor young ladies who do not want to return to Torneå, and to whom I would very much like to be of service and to be good for something'.[16] But at the same time the sisters showed up in Paris, a gushing romantic song that de Maupertuis had composed in adoration to one of them leaked out. While not denying that he had written the paean, de Maupertuis tried to deflect attention from his relationship with the two women, who had already become objects of considerable curiosity in Paris society. In a letter to his own sister, Marie, in January 1739 de Maupertuis referred to them as 'young Swedish ladies' and sought to divert responsibility for any tryst or attachment onto the Arctic expedition's artist, Herbelot. He wrote:

> Our artist had made one of them sacred promises of marriage and told so many lies about his wealth, saying he was very rich and could make her a great lady, that this poor girl came to find him, and her sister, counting on sharing this imaginary fortune, accompanied her. Not only is this artist as poor as a church mouse, but he was unfaithful and married someone else.[17]

He went on:

> All the company of the polar circle have contributed [to a fund for their support] and we have placed [the sisters] in a Paris convent to learn French and develop a taste, if possible, for our religion. I was quite astonished by the desire of these young ladies to come to France, not knowing at all that the youngest was coming there to marry the artist, nor anything about the promises he had made. It is quite embarrassing for us to have to find situations for two young ladies.[18]

The mischievous Voltaire again poked fun at de Maupertuis and the consequences of him and his companions being so free with their affections while in Lapland. He composed a verse about the ongoing debate over the shape of the Earth, including the lines: 'Bring back from the climes governed by Sweden / Your rulers, your sectors, and above all two Lapp women.'[19]

Cassini and his supporters pounced upon the Planström sisters' arrival and sought to extract maximum humiliation from their presence to undermine the personnel, behaviour and science of the polar expedition. Desperate to wriggle free from the accusations and scurrilous chatter, de Maupertuis considered sending Christina and Elisabeth to live with his sister in Saint-Malo. But he decided, probably rightly, that this would just create and attach more scandal to him in his hometown. So instead, once the sisters had renounced Lutheranism, he organised small pensions for both and separated them.

Christina entered a convent permanently while, with a dowry from Duchesse d'Aiguillon (who was at the time rumoured to be de Maupertuis' lover), Elisabeth married a minor nobleman. The marriage did not turn out well – her reprobate and possibly bigamous husband abused and then abandoned her, attempting to abscond with the dowry. It was a sad and ignoble end to a long journey spurred by the promise of love.

△△

To seize back the initiative, in summer 1738 de Maupertuis finally published his own record of the Arctic expedition and its conclusions. *La Figure de la Terre*[20] took a typically bravura stance. The book depicted its author as part adventurer, part scientist and part statesman, a courageous, principled man who had conquered perils to follow his convictions and prove the validity and superiority of Newton's theories. De Maupertuis quickly translated the work into English and sent copies to Bradley, Graham and Machin in London to seek their backing.

But he knew that words alone would not settle the debate; they needed extra, irrefutable evidence to back up the Arctic findings. For this, de Maupertuis turned again to Celsius. He suggested that the two of them secretly carry out another arc meridian measurement on the ice of Lake Vättern in south-central Sweden, using the Graham sector to verify their results. He wrote:

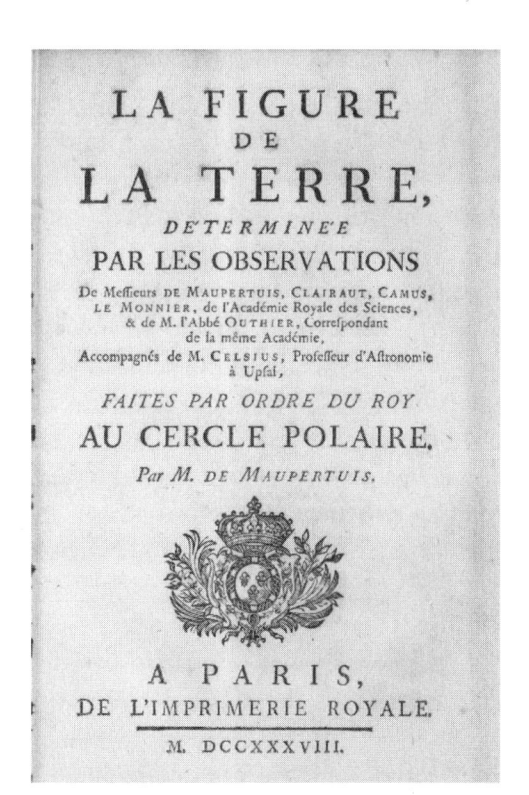

De Maupertuis' 1738 account of the Arctic expedition and its findings to establish the shape of the Earth, *La Figure de la Terre*.

I am sure that if we undertake [the survey], it will still be measured before the return of those from Peru. I will be able to go to Gottenbourg this Autumn and we will be able to measure this lake this Winter either by the pole or by chain; and as we will be able to use unintelligent people in this work where it is almost only a question of knowing how to count well, I have no doubt that the measurement cannot be made in 15 or 20 days.

But I beg you above all that the thing remains secret between us until it has succeeded; because after the way things have turned out, I am very much afraid of going to carry out an announced undertaking which would not succeed. No one after the discontents I have had will be surprised to see me return to a place that people know I love and that I haven't seen enough.

The work of which I speak will be all the more agreeable to me as it procures me still the pleasure of living with you.[21]

Whether Celsius was fatigued after his long travels, or he had genuine concerns for the plan's safety, or simply felt disinclined to leave his hard-earned professorship again now that he achieved financial security, he balked at the suggestion and resisted de Maupertuis' excited overture. The ice, he argued, could not be relied upon, and he was unwilling to go along with the scheme.

This moment marked a profound and permanent cooling in relations between the two men. After all they had endured and achieved together, de Maupertuis was bitterly disappointed that his Swedish friend was so reluctant to help him, even in his own country. At the same time, it seems that Celsius was waking up to the Frenchman's devious ways. He had already acted against type to defend de Maupertuis so vigorously, and was apparently reluctant to put himself in the firing line again. Celsius was not forthcoming with any alternatives to the Lake Vättern proposal. Five years after leaving for his European tour, he was happily back with his family and colleagues in the familiar surroundings of Uppsala. And his mind was turning to other subjects.

De Maupertuis was rebuffed but not discouraged. In summer 1739, he decided to take the battle directly into Cassini's territory by remeasuring some of Picard's original meridian between the cathedral in Amiens and Notre Dame Cathedral in Paris. He reassembled key members of the

Arctic group – Camus, Clairaut and Le Monnier – and they set out with the Graham sector to challenge the figures underpinning Cassini's belief in a prolate planet squeezed at the equator.

De Maupertuis simultaneously attempted to position this venture as being of wider use and application: 'Independent of the utility of this measure for determining the shape of the Earth, it will also be of considerable utility for the particular geography of France', he wrote.[22] Might this possibly have been intended as an olive branch to Cassini? Or was it merely another dig at his opponents and their focus on producing a definitive map of their homeland?

To burnish his credentials as the all-action hero, and in particular, to curry favour among sympathetic women of the aristocracy, de Maupertuis also commissioned a portrait of himself. Over two decades before, he had sat for a family portrait by the acclaimed artist Robert Le Vrac De Tournières, and in autumn 1739 he presented himself to the painter again. He was staking his claim for a place in the illustrious company of politicians and nobility.

The resulting portrait was a flamboyant piece of self-advertisement. It shows de Maupertuis in an improbably luxurious Lapp hat and outfit, half-sitting at a window in front of an idyllic Arctic scene, with smoke from mountain huts billowing out into the freezing air. Around him are strewn maps of the scientists' triangulations and a pair of the fur leggings they had worn against the extreme cold. With his left hand, de Maupertuis pushes down onto the top of a globe – compressing it slightly like a Dutch cheese into its now proven oblate form.

With sleight of hand and the painter's brush, the portrait took the collaborative success of the Lapland expedition and attached it to de Maupertuis as his personal triumph. On seeing the painting, Voltaire commented that its subject had flattened 'both the Earth and the Cassinis'.[23] But the battle still was not over.

With the equatorial expedition still yet to return and formally report, l'Académie kept up a studious neutrality, loath to come down decisively in favour of either side. This prompted a despairing de Maupertuis to deploy ever more bizarre tactics in support of his cause. First, he decamped back home to Saint-Malo, publicly complaining of malicious and unfair treatment by his peers. Then in autumn 1739 he joined the royal court during its annual two-month sojourn at Chateau Fontainebleau, south-east of

A heroic depiction of the adventurer-scientist de Maupertuis after his triumphant return from the Arctic – his hand depressing the globe into its proven shape and a fanciful Lapland scene behind him. Portrait by Robert Le Vrac De Tournières (1667–1752).

Paris. Here, amid the splendour recently expanded to accommodate Louis XV's ever-growing retinue, Minister Maurepas persuaded him to accept a new honorific position. The purpose of the role was to 'perfect navigation', and it came with the added sweetener of a substantial pension of 3,000 livres per year.

Another of de Maupertuis' high-society female admirers, Duchesse de Saint-Pierre, wrote to congratulate him: 'It's a new life for you, in the tumult of the great people at court, becoming a gambler, and even a hunter. What a life for a mathematician! You are extremely à la mode. Think of your affairs and profit from your long stay.'[24]

De Maupertuis was clearly alive to the honour and advantage conferred upon him. 'It is a position created expressly for me by M. de Maurepas. Everything was done in the most gracious way for me. It makes me very happy and quite comfortable', he wrote in December 1739.[25] But nonetheless, he continued plotting and seeking ways to clinch a final victory.

The following year, de Maupertuis embarked upon an elaborate literary hoax by producing an anonymous book, *A Disinterested Examination of the Different Works on the Shape of the Earth*, bearing the fictitious publication date of 1738 and place of origin, 'Oldenbourg'.[26] The volume purported to examine the debate about the shape of the Earth from an impartial standpoint, beginning with an overview of the competing surveys. Five pages about the Arctic expedition were followed by twenty-eight pages on the measurements of Picard and the Cassinis, packed with subtle but telling criticism. While working hard to disavow and keep the book's authorship secret, de Maupertuis made sure it reached those across Europe who he knew would take notice. Via the Royal Society in London, he sent copies to Folkes, Jurin, Bradley and Desaguliers, and also to Bernoulli and König in Switzerland.

One of Cassini's closest associates, the astronomer, geophysicist and chronobiologist Jean-Jacques d'Ortous de Mairan, praised the book's tone and erudition: 'Whatever [the identity of the author] may be, this book is a credit to whoever wrote it. It is in the hands of everyone in Paris, including the ladies, who also dabble in wanting to know the shape of the planet they walk upon.'[27]

Others, like Voltaire and Algarotti, immediately got the joke and surmised who was behind it. But the press turned the book into a minor literary phenomenon by conjecturing upon the author's identity and motives. De Maupertuis later ratcheted up the pressure by issuing a second edition (again falsely dated), which included a preface covering 'the History of this book', with further contradictory and misleading trails

De Maupertuis' anonymous satire *A Disinterested Examination of the Different Works on the Shape of the Earth* (1740). It cleverly destabilised the Cassinis while keeping the author firmly in the public eye.

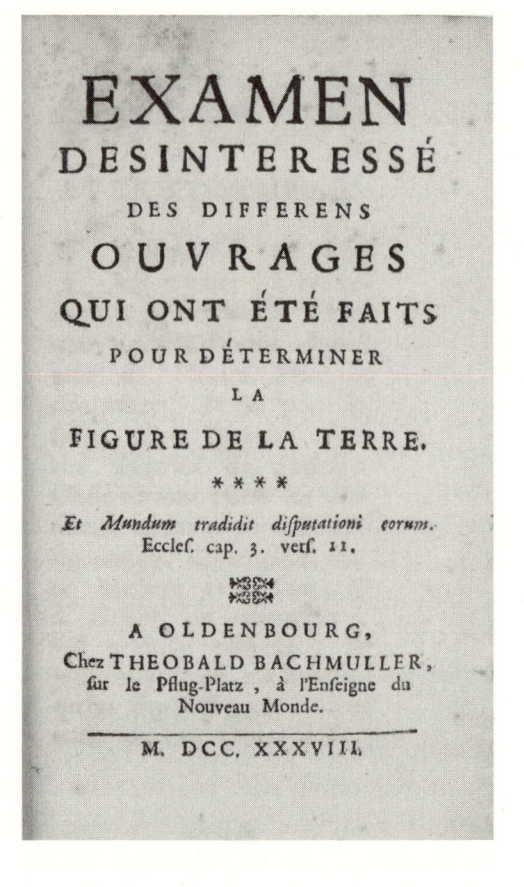

about its provenance. The preface even quoted Cassini's largely positive response to the first edition, boasting:

> This book itself became an object of interest for that type of connoisseur who prides himself on knowing everything about literature, from the ink to the style. People in high society took a different interest in it, since it speaks continually of the savants with whom they live, whom they esteem or scorn.[28]

This scheming allowed de Maupertuis to demean his opponents, while displaying his clever wit to those who understood the ruse. It also shifted the debate beyond science and l'Académie to conversations in literary salons, newspaper gossip and out into the wider republic of letters. It was an effective if underhand strategy; de Maupertuis could not have risked his position by penning such an attack on a fellow academician in his own name. He wanted to defend his reputation as a mathematician and build broader social credibility.

Ultimately, and after so many years of disagreement, it was not the efforts of de Maupertuis, Celsius or any of their fellow Newtonians that settled the question over the shape of the Earth, but the Cassini camp itself. At the same time as de Maupertuis and his Arctic veterans were checking and challenging Picard's measurements, the youngest Cassini also decided to revisit his father's and grandfather's work. César-François Cassini de Thury was just 25 years old, equipped with an agile mind and a flair for rhetoric that outshone even his forebears. He brought a fresh and open perspective, unhindered by the tenets of previous generations.

In April 1740, Cassini de Thury presented his findings to l'Académie. He acknowledged errors in his predecessors' measurements in France and expressed the likelihood that the Arctic conclusions about the flattening of the Earth were correct.[29] However, his surrender – if that is what it was – was hedged with caveats and qualifications: 'It is extremely difficult to avoid little errors, which when multiplied may produce quite considerable ones over an extent as large as this line.'[30] Although he also recognised the Lapland measures as having 'all the exactitude that one could wish [to demonstrate that] degrees [of latitude] increase in length as they approach the poles and we are persuaded that they will be confirmed by those currently underway in Peru'.[31]

While acquiescing to the Earth being flattened at its poles, de Thury skilfully presented this as being secondary to his family's principal object of completing the map of France. But de Maupertuis recognised the significance of the apparent withdrawal. In another letter to Bernoulli, he wrote: 'The Cassinis have publicly recanted on the elongation of the earth and have confessed to having found it flattened in the sixth operation that they have just finished. They are a bit the talk of the town and well deserve it.'[32]

Various remeasurements and technological breakthroughs in the nineteenth and twentieth centuries showed that, although Newton's theories had triumphed overall, the detailed findings from the 1736–37 Arctic expedition were not exactly correct. It was later discovered that the Earth is composed of both solid elements and liquid layers in constant motion, rather than being a homogenous planet of uniform density. It also became apparent that the astronomers' observations and calculations in Lapland

César-François Cassini de Thury (1714–84) – the third generation of the Cassini family to head the Paris Observatory, and the first to accede to Newton's universal theory of gravitation.

had been distorted by various atmospheric effects and qualities of light. But by the time La Condamine returned from South America to France in 1745 and eventually published his own expedition journal in 1751,[33] the science was beyond question. De Maupertuis, Celsius and their colleagues had proved that the Earth is a spheroid flattened at its poles.

The decade-long dispute between the two sides has strong parallels with the modern controversy over global warming and climate change. Today's vested interests are just as powerful, and the debate every bit as heated, as in the first half of the 1700s. And the tactics deployed by those keen to discredit their opponents are similarly vicious and deceitful. Eighteenth-century pamphleteering and reputation wars had just as much bile as twenty-first-century social media spats. And then, as now, facts and evidence on their own remained bit-players in the struggle to command public acceptance and opinion.

In the 1730s at least, truth and reason finally won out. But for Celsius, although he had been vindicated, the whole saga had become steadily more wearisome and distracting. He was eager to pick up his studies and direct his increased standing and resources to different priorities. Even when writing the 1738 defence of the expedition's work in the Arctic, its impact upon him was apparent. Sending the pamphlet to Cromwell Mortimer at the Royal Society in London on 14 March that year, he wrote: 'Your last letter found me at Uppsala, after the departure of the French gentlemen. I have been in a bad state of health after my return hither, and had a violent fever at the beginning of this year.'[34]

Celsius could not have been sure, but might have suspected: he only had a few years left to continue his science and leave his mark on the world.

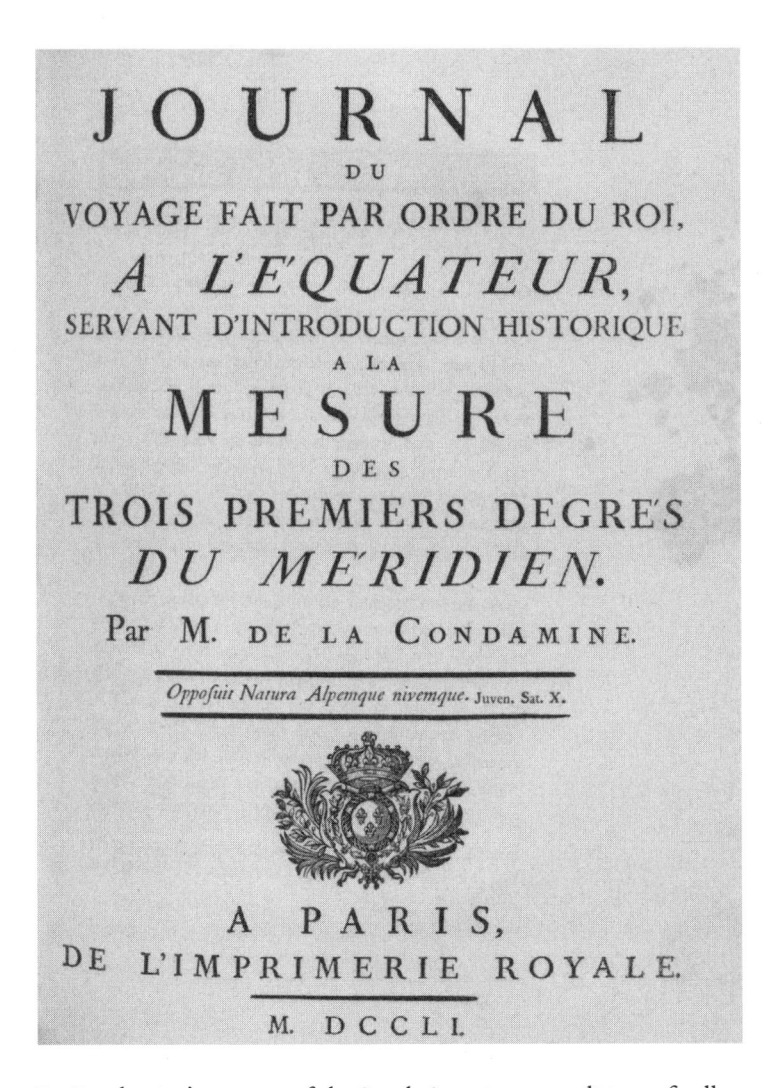

JOURNAL
DU
VOYAGE FAIT PAR ORDRE DU ROI,
A L'ÉQUATEUR,
SERVANT D'INTRODUCTION HISTORIQUE
A LA
MESURE
DES
TROIS PREMIERS DEGRÉS
DU MÉRIDIEN.
Par M. DE LA CONDAMINE.

Opposuit Natura Alpemque nivemque. Juven. Sat. X.

A PARIS,
DE L'IMPRIMERIE ROYALE.

M. DCCLI.

La Condamine's account of the South American expedition – finally published sixteen years after his party had set out, by which time the question they aimed to resolve had long been settled.

PART IV

SEA AND SPACE

16

SERVING AND OBSERVING

Uppsala's First Observatory (1741)

In the field of observation, chance favours only the prepared mind.
Louis Pasteur, speech to the Faculty of Sciences,
Lille University, 7 December 1854

Apocryphal or otherwise, stories like the apple falling on Newton's head or Archimedes leaping from his bath with a cry of 'Eureka!' illustrate the central role of observation in science. Depending where in the continuous, cyclical process of the scientific method one begins, observation is often seen as the catalyst or first step to discovery. What researchers observe – be it directly through their own senses or with the most sophisticated instruments, equipment and models – leads them onto questions, hypotheses, experiments and conclusions. Observation generates data, which can then yield evidence, understanding and wisdom.

It was in this spirit that Celsius returned to Uppsala in summer 1737, ready to pick up his professorial duties and determined to create an observatory to emulate those he had seen in Germany, Italy, France and England. In Berlin, five years before, Celsius had noted with envy in his tour journal the Kirch family's financial arrangements:

> I would be so happy, when I, if God wills, come home from my travels, I too could get privileges for almanacs; so that our society could get a salary for its secretary, and moreover I got funds to build an observatory. I am now making myself in every way, as much as I can, aware of all the issues, which should be taken into account in constructing an observatory.[1]

In the time he had spent away, Celsius had transitioned from fresh-faced promise to full confidence and maturity. A portrait, completed around 1740, shows him illuminated against darkness, now rather broader of figure, wearing a grey, tailored coat, with a matching waistcoat and billowing white shirt tied tight at the neck. His right hand rests casually at his waist, the assured stance of someone safely returned from the defining exploits of his life.

Celsius' expression suggests someone whose standing is enhanced and their significance secured. The face is still benign, but more weathered and timeworn. The person staring out from the canvas is a deeply learned man of stature. He has the look of someone who knows things about the past, present and future, and who is determined to put this knowledge to practical use.

With no independent home of his own in the city, Celsius moved in with his mother and sister at their eating house alongside the river in Svartbäcksgatan. After so long on the road and all the energy-sapping demands of the Arctic, it must have felt safe and familiar. With his energies restored, he set about building an observatory in the garden. This would have been a familiar sight to Gunilla – the daughter and wife of astronomers – who watched on as her son erected a modest outbuilding and filled it with instruments to continue his father's and grandfather's tradition of home-based stargazing.

Evidently pleased with the result, Celsius wrote to Benzelius:

> I have now, since my return home to Uppsala, had a small observatory built in my garden. Here I have put up the 3-foot quadrant bought in Paris. I also have a good clock, tubes of considerable length and a tube with a micrometer by Mr Graham in London. With this, I can perform all astronomical observations quite well, as they do in Paris. The difference is only that the observatory in Paris is one of the most splendid palaces in Europe, and my own is just a small wooden hut.

With his private observatory complete, and the financial security of his pension from the French King behind him, Celsius turned to his responsibilities as professor of astronomy and secretary to the Swedish Royal Society of Sciences. The state of the latter was disappointing; while Celsius had been away, his temporary replacement, Samuel Klingenstierna, had done little to advance the profile or work of the society. It had published no scientific papers at all during Celsius' absence, so he found a pile of unreviewed submissions awaiting his attention. With typical gusto,

Celsius attacked the task, reading, reviewing and publishing dozens of the neglected papers, and soon adding some of his own.

The university situation was much better. Despite not having an academic degree himself, Olof Hiorter had filled Celsius' shoes admirably. He had given all the necessary lectures and even travelled to observe a total solar eclipse in southern Sweden. He had shown both the acuity and appetite for heavyweight mathematical calculations. Celsius was keen to delegate this sort of intellectual labour, so decided to engage the 41-year-old Hiorter as his assistant, thereby freeing up his own time for more original and theoretical pursuits. It was an appointment that brought professional and personal rewards to both men.

Hiorter had eaten and enjoyed the atmosphere at Gunilla's restaurant while he was standing in for Celsius, but with the professor now back and his assistant role secured, he became a regular visitor. And welcomed and accepted into the family scene alongside Celsius, Hiorter fell in love with and then married Celsius' sister, Sara-Märta.

This union brought with it another advantage: Hiorter introduced Celsius to a gifted student, Pehr Wargentin, who he thought might be able to support their future work together. Aged just 20, Wargentin was a vicar's son from Jämtland, the central western province of Sweden bordering Norway. Despite his faith and ministry, Wargentin's father had encouraged his son to take an interest in science from an early age – a fascination that burst into life when he witnessed a total eclipse of the Sun as a 12-year-old. Wargentin came to Uppsala University in 1735 and quickly demonstrated a clear and penetrating mind, backed up by boundless energy and strong moral integrity.

Celsius possibly recognised some of his own qualities in Wargentin. He decided to appoint the student, who already displayed a special talent for statistical analysis, as a further assistant. The core of the team that would support the rest of Celsius' career was now in place. And with it grew the desire to think bigger and more imaginatively.

Celsius' able stand-in as professor, and later his brother-in-law, Olof Hiorter.

While making astronomical observations in rough wooden huts on Arctic hilltops had been a necessity, Celsius could never be content with this at home. He wanted a real observatory, designed for and dedicated to his particular interests.

☾

In 1738, Celsius started a campaign to persuade the University authorities (including the Senate, of which, as a senior academic, he was a member) of the case for a purpose-built observatory in Uppsala. Having watched the master manipulator de Maupertuis' wily ways of lobbying influential people, Celsius began to apply some of the same tactics.

That year saw a sea-change in Swedish politics, with the Caps Party, which had held control of the country's parliament for almost two decades, being replaced by the Hattar (Hats) Party. The groups' names indicated both their characteristics and supporter bases. The Hats, named after the pointed tricorn headgear of officers and gentlemen, were muscular and assertive, especially in foreign affairs. They favoured an alliance with France and were keen to seize back from Russia the Baltic territories lost during the Great Northern War at the beginning of the century. By comparison, the Caps were viewed as soft and timid – mainly representing the clergy and peasants.

Celsius saw an opportunity to ride this wave of change by making direct contact with the Hats leader, Carl Gyllengborg, who was also, by happy coincidence, the chancellor of Uppsala University. On 8 June 1738 he sent a lengthy letter reminding Gyllengborg that proposals to create an observatory in the city, including the idea of refurbishing one of the castle's fire-damaged towers to fulfil this purpose, had been discussed for decades. Drawing on his international reputation and playing skilfully to the chancellor's ego, he went on: 'It is well known that in all well-established kingdoms, even those we here call Barbarians, astronomical observatories are built at public expense.'

The gambit worked. Just two days later, the Senate approved 9,000 copper daler (around five times Celsius' annual salary) for building costs, once a suitable site was found. By 24 August, Celsius had the answer – a substantial, three-storey *stenhuset* (stone house) at 9b Svartbäcksgatan, just a few doors down from his mother's house, which he declared could be readily converted into an observatory. The Senate obliged once again, agreeing a further 8,000 daler to cover the purchase of the property.

With this funding in place, Celsius approached Carl Hårleman, a well-known architect and court superintendent, who was already busy on a number of other building projects in the city. By another twist of good fortune, Hårleman (just a year older than Celsius) had also been one of Hiorter's students, so he had some insight into what the observatory would provide and do. The architect produced detailed plans and elevation drawings to show how the house could be converted into a state-of-the-art scientific amenity.

Hårleman proposed that a new rectangular viewing tower be constructed on the building's flat roof, 16 cubits (9.5 metres) by 14 cubits (8.3 metres) in area, with three full-height (7 cubits, 4.2 metres) narrow windows on each face to permit observations in every direction. Below, on the building's existing floors, he included a lecture theatre, library and sleeping accommodation for the professor, plus more rooms and working spaces for his assistants. To announce the building's new function, the whole edifice would be topped by an iron celestial sphere – the Earth in the middle, surrounded by planets and stars. The observatory design lacked the magnificent scale of its counterpart in Paris and the commanding position of Greenwich, but it was a confident sign of intent. Celsius wanted his observatory to be Earth-centred and focused firmly on practical application.

Architect of Celsius' observatory in Uppsala and later the Stockholm Observatory, Carl Hårleman (1700–53).

On 12 April 1739, Celsius signed a contract with stonemason and builder Johan Christopher Körner to begin all the work necessary to realise Hårleman's plan. The document listed all the quantities and specifications of lime, plaster, carpentry, glazing and other trades and materials, with details of how Körner was to operate and subcontract. It was a construction agreement written with the eye of an empirical scientist, which concluded with the sum of 11,500 daler. Although this amount was more than the Senate had originally approved for the refurbishment, they waved it through – presumably feeling they had little choice since the builder had already started the job.

Mindful of this overspend and conscious that other costs might occur, Celsius kept up the campaign for more funds to equip, furnish and

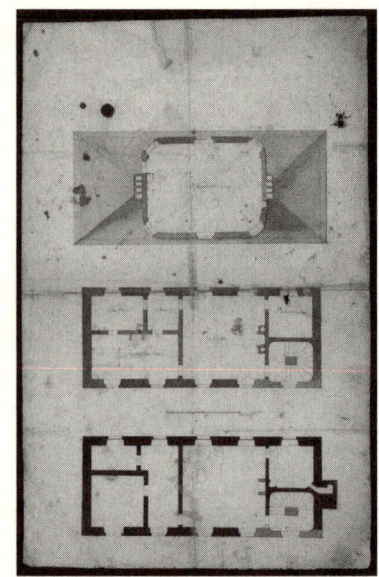

Hårleman's architectural designs for the Uppsala observatory.

open the observatory. On 30 October, he published a virtuoso pamphlet entitled *The Benefits of an Astronomical Observatory in Sweden*.[2] It opened with the sweeping proclamation: 'It would be easy to prove the usefulness and necessity of a Swedish Observatory, if one were content to cite only all the examples of modern evidence as examples.'

The pamphlet set out four core objectives for the observatory to support. The first was to improve navigational methods, including the then unresolved problem of reliably determining longitude at sea. Secondly, Celsius promised that the new facility would support better geographical mapping of the Swedish kingdom – a project to rival that of the Cassinis in France. The third target was greater accuracy of timing, calendars and almanacs, including discrepancies between the papally decreed versions used in Catholic countries across Europe and those in Protestant states like Sweden. And finally, Celsius explained how the observatory would allow him to return to the study of weather patterns, which he and Burman had begun just a few streets away, seventeen years before.

To back up each of these goals, Celsius painted patriotic pictures of their usefulness. Reaching back into his Uppland roots and history, he wrote: 'Truly no place in the whole of Sweden can be found more convenient for this than Uppsala, the ancient capital of the entire Nordic region, the Swedish Church Archbishop's See, the oldest mother of all Swedish studies, located in the core and heart of the Svea forest.'³ And with his usual eye to the long-term future, he explained why more accurate calendars were needed: 'Correction so wide is necessary as would happen that in 10,000 years hereafter the vernal equinox would fall in December, and Easter come at Christmas time.'⁴

Celsius undertook to 'complete a catalogue of the fixed stars of the Zodiac' and produce new and better depictions of the Sun, Moon and Jupiter. He also added a personal touch, saying: 'Nothing should be more important to me than this observatory.'⁵

It was a clever and persuasive piece of writing, which drew upon every aspect of his character and international status. And, once again, it got results. The Senate signed off another 5,000 daler to pay for a pendulum clock and giant zenith sector (a copy of that used in Lapland) from George Graham in London. It was only after this decision that Celsius revealed these items had already been made and had arrived by ship in Stockholm. He had gambled on the Senate's acquiescence by secretly sending a local instrument maker, Daniel Ekström, to commission and collect them from his craftsman friend in Fleet Street.

Through a series of deft, incremental moves worthy of de Maupertuis' manoeuvrings in France, Celsius had gained the University's trust and convinced its governing body to all but empty its coffers. With a final bill of more than 33,000 daler, the observatory was finished and opened in summer 1741. Everything now depended on it and its team.

☾

The Ångströmlaboratoriet lies a few kilometres outside the centre of Uppsala. It is the modern equivalent of Celsius' academic base. I went there in the pleasant and helpful company of Dr Eric Stempels, researcher at the University's Department of Physics and Astronomy. Originally from the Netherlands, Eric is one of a trio of contemporary experts whose roles, specialisms and responsibilities most closely match those of Celsius. He is the author of a book about Celsius' father Nils, including the first translation into Swedish of his controversial 1679 dissertation. Eric has an impressive command of the family's deep footprints in science and Uppsala. Like me, he clearly finds the Celsius story both captivating and acutely relevant to the modern world.

We entered the red, five-storey building, which is topped with a substantial observatory dome. In the main staircase hangs the brass tube of the Graham sector that Celsius so cannily ordered in 1740. Up close, the size of this instrument becomes even more apparent, its thick brass tube extending vertically through two whole floors. As I photographed Eric standing next to it for scale, my mind went back to the heavy wooden tripod that once supported the tube and the struggle involved in hauling its companion instrument up and down mountains in Lapland.

Eric showed me into a neat meeting room named after and dedicated to the three generations of Celsius astronomers. The 1730 portrait of Anders by Olof Arenius is flanked by the likenesses of Magnus and Nils Celsius, above which hangs the massive dark wood casing of one of the observatory's original telescopes. In one corner is the Graham clock. Eric carefully wound it with a heavy brass key and then set the pendulum swinging. Most impressive of all, though, was the enormous document and instrument cupboard standing opposite – almost 2.5 metres square of freestanding polished wood, still stacked with papers and books. It is the only piece of furniture that survives from the observatory.

Further along the corridor, past more gilt-framed portraits of Anders Spole, Pehr Elvius, Samuel Klingenstierna and other key people in Celsius' life, Eric tapped a series of security codes into a keypad and took us into really hallowed territory. Here was Celsius' library, with the thousands of books, papers and artefacts he amassed or inherited from his grandfathers and father, all bequeathed to the University after his death. The books, many bound in crackly white vellum with embossed gold titles, sit within neat, custom-made, locked glass cases, flanked by a pair of terrestrial and celestial standing globes from around 1760. Eric leaned over the latter and pointed out something unusual on the top – an arrangement of stars marked in the shape of a reindeer. It was a quaint Scandinavian relic of the days before the International Astronomical Union standardised the composition and boundaries of eighty-eight constellations in the 1920s.[6]

A door inside the library led, via yet more electronic security, to the most precious objects of all, lining a windowless space the size of a small bedroom. In the centre stood Anders Spole's deep-carved, multicoloured valuables box, which he used on his 1695 trip to the Torne Valley. And, from one of the shelves, Eric pulled out the stained oblong case of the compass-cum-sundial-cum-quadrant that Celsius took to the Arctic in 1736. The brass components have disappeared, but the ivory compass dial inside the hinged lid remains bright and spectacular, painted with an unsettling array of skulls, skeletons and the puffed faces of the classical south, east and west winds – *Notos*, *Eurus* and *Zephyrus*. We both went quiet, appreciating the importance of this instrument. It is an object that still seems to hum with the connection between its owner and the natural world.

17

A VAST, PROFOUND TRUTH

Unravelling the Baltic Sea Mystery (1741–43)

Everything we see is perspective, not the truth.
Attrib. Marcus Aurelius Antoninus (121–80 CE)

The grey seal bobbed patiently in the water, waiting for a gentle swell to lift her towards the granite hump on the Baltic shoreline. She had fixed the green-black obsidian orbs of her eyes on this spot and was determined to make it her own. The moment came: a fine-foamed wave washed into the curve of the bay, propelling her headfirst to the edge. The blunt claws of her foreflippers grasped the rock. And with a series of rippling, full-body convulsions, she hauled the 180 kilograms of her bulk onto its surface. Each blubbery shunt was harder than the last as the sea released her from its buoyancy, but each more confident as she gained traction and the sanctuary she sought.

With a belly full after feasting on sand eels, sprats and herring, she was regaining weight and strength now that she could forage again – her most recent pup weaned and gone away. She turned her neck right then left and raised her snout, sniffing the air, her whiskers twitching and beaded with droplets. It was time for rest.

Her fur soon began to dry in the fresh breeze and sunshine, the mottled black and brown patches along her back losing their sheen as water turned to air. She half-rolled over, exposing her white underside, one flipper pointing skyward, as sleep drifted over her. The stone she lay upon would change the way humans viewed the world.

Two hundred years later, Anders Celsius stood alone at the same spot. It was summer 1731. He had travelled to the tiny islet of Lövgrund – 100 kilometres north of Uppsala and a stomach-heaving boat ride away from Sweden's east coast. His salt-crusted boots searched for grip on the rock as he gazed down at the clear water lapping almost a metre below. No marine mammal, however deft or determined, could now possibly climb so high. What could have caused such a change between sea and land?

Almost three centuries after Celsius, I also came to this place, to get a sense of what he saw and heard and witness this example of Earth's dynamism. In the tiny hamlet of wooden cottages that fringed the cove I saw an old boathouse, now perched high and inaccessible above the water, and striated lines of different-coloured pebbles on the beach, marking where the sea had once reached. A swan dipped its head below the surface in search of food and a gaggle of geese streamed overhead. It was a peaceful spot, but not silent. The wind stirred the trees into a constant background whisper. The sound was a reminder that this is a landscape in motion and transition, an environment where elemental forces collide.

By the time Celsius stood here, the mysterious phenomenon of the Baltic Sea's apparent steady retreat had been known about for centuries. Generations of the region's sailors, fishermen and women fought a constant losing battle with harbours that dried up, inlets they could no longer navigate and creeks that led nowhere.

Carvings on the rune stone at Runby – beside the sea 1,000 years ago, but now far inland.

A poignant record of the problem still survives from the Viking era, carved into a rune stone in Runby, a former coastal village now subsumed into Stockholm's north-western suburbs. The stone was erected around 1050 by a local woman, Ingrid, recording her construction of a *laðbro* (literally 'loading bridge', quay or wharf). Translated into English, a carved ribbon twisting around the mythical animals on one side reads: 'Ingrid had the loading bridge made and the stone carved after Ingemar, her husband, and after Dan and after Banke, her sons. They lived in Runby and there owned a farm.' And on the other side of the dun erratic boulder deposited there by the last ice age she wrote: 'This shall stand as a memorial as long as humans live.'

Ingrid's wish has lasted for a millennium, but the sea for which the *laðbro* was built is now several kilometres away, at least 5 metres below its level in her time. And the stone that once stood at the edge of a sheltered anchorage on one of the main Viking sailing routes now looks out on a quiet peripheral road and the overhead lines of the railway leading to Stockholm.

Celsius had also heard about Östhammar – a once thriving port southeast of where he stood. In 1491, its population had moved the whole town so they could stay connected to the receding shore. That year, four of Östhammar's most prominent citizens, Anders Botvidsson, Nils Hansson, Anders Persson and Jöns Andersson, formed a delegation to the Swedish government in Uppsala. Its mission was to complain and seek help about their dockside no longer being accessible by boat.

The four men, dressed in their Sunday best, entered the unfamiliar city with trepidation. By good fortune they caught the sympathetic ear of Uppsala's archbishop, Jacob Ulfsson. At this time, Sweden was part of a single kingdom that also included Denmark and Norway, with a common sovereign based in Copenhagen. Ulfsson was his representative (effectively prime minister) in Sweden.

Upon hearing the townsmen's plight, he issued a resolution, beginning with a vivid summary of their difficulties:

> During recent years the land has grown outside the town at the sea, so that where some years ago a cargo boat of five or six Swedish läster [about 15 tons] could come from the sea into the town of Östhammar, not even a fishing-boat can go nowadays. And the land is still growing and rising every year.[1]

The document granted the petitioners the permission they sought: to abandon the town and build a replacement, Öregrund, with a brand-new harbour, church and houses.

Celsius had also heard another similar, but more recent, tale about the coastal town of Luleå, further north up into the Gulf of Bothnia. Linnaeus visited there on his trip to Lapland in 1732, and it was where Meldercreutz and Lord Cederström intercepted de Maupertuis' expedition on its way north four years later. By a 1648 royal decree of Queen Christina, Luleå had also been repositioned because it had become too distant from the water. What on Earth was causing all this inconvenience, upheaval and expense?

≈

Others had searched for answers before Celsius. In 1706, the high-powered duo of Urban Hiärne, a Swedish scientist specialising in chemistry and medicine, and Elias Brenner, a historian born and raised on the Finnish side of the Baltic, embarked on the first systematic study of what was happening to the sea and land.

As the father of twenty-six children, Hiärne had more reason than most for seeking to comprehend and explain the world around him. In a pioneering version of mass observation and crowd-sourcing, he sent a questionnaire to bishops, governors and other state officials across Sweden and Finland asking them for evidence and opinions about a range of natural phenomena, especially trends they had witnessed over time. From the dozens of written responses, some lengthy and detailed, Hiärne concluded that the sea level in the Gulf of Bothnia – most probably the whole of the Baltic – was falling. He wrote:

> From old times one has noticed and experienced how large bays of the sea gradually turn into land. Where people in several places seventy or eighty years ago could sail freely with small boats, now every year hundreds of loads of hay are gathered. One has also found that the waters up to one Swedish mile [around 10 kilometres] or more off the coast gradually become more unsafe. This has been experienced by several navigators whose ships have run aground and been damaged, where never before any cliffs or grounds have been detected.[2]

So the issue had been ascribed to both the land rising (Ulfsson) and to the sea falling (Hiärne). The latter speculated that the hundreds of rivers feeding into the narrow trough of the Baltic must originally have made the sea level much higher, with gravity over the centuries slowly emptying this natural header tank into the North Sea. Maybe it was a final ebb of the biblical flood, with the Baltic still draining through its southern outlet. Or was some other unimaginable, unseen force sucking away the water and causing the sea to disappear?

Celsius gazed at the beach of jumbled stones at Lövgrund and contemplated the nature and form of Uppland's landscape. In this part of Scandinavia at least, perhaps the Earth's surface was insidiously rising. If so, why and how quickly was this happening? For how long could this go on or when might it stop? Was the youthful Earth still growing? Whatever the truth, he needed to study, discover, understand and explain.

≈

Celsius had also pinpointed a second 'seal rock', a short distance north at another small island, Iggön, which he engaged a local mathematics teacher, Johan Rudman, to reconnoitre. Celsius unearthed a tax certificate from 1583, sent to two brothers who then owned this small, wooded skerry. Their late father Nils had been dubbed 'Rik [rich] Nils' because of the abundant fishing and hunting he enjoyed there.

The conspicuous rock at Iggön was a favourite location for Nils to steal up on and harpoon resting seals. But as its top and the water drew further apart, fewer seals came, so the resourceful hunter took action. He lit a fire of brushwood and thick pine logs on top of the rock, causing it to fracture along a lower, horizontal plane and so creating a new place for seals to come ashore and within his sights again. Nils was no scientist, but he had unknowingly formed an outdoor laboratory, where Celsius could now examine and gauge the rate of land and sea change.

Rudman made a sketch map of Iggön with the burnt and split stone clearly marked. And at Lövgrund, on Celsius' instruction, he also carved a short horizontal line and the date '1731' into the seal rock there. This

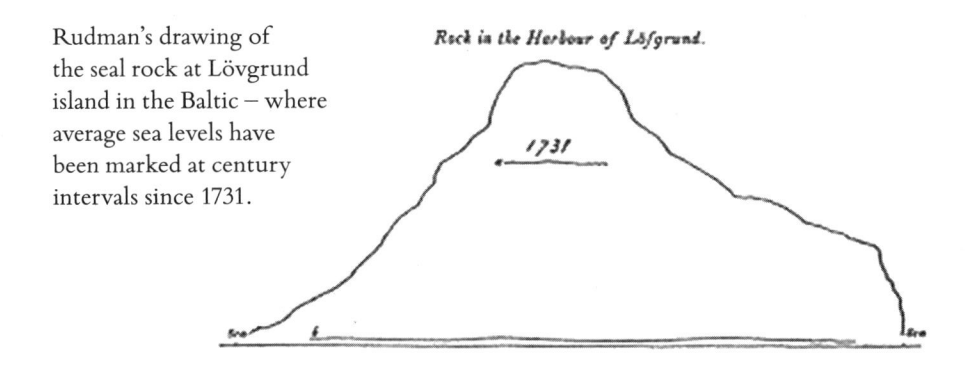

Rudman's drawing of the seal rock at Lövgrund island in the Baltic – where average sea levels have been marked at century intervals since 1731.

indicated what he estimated as mean sea level, allowing for winds and changes in the season. The Baltic baseline was set.

Iggön is no longer an island. Today, it forms a finger-shaped peninsula jutting eastwards into the sea. Rik Nils' rock is still there, but surrounded now by dense trees, high and dry 15 metres inland and nearly 3 metres above sea level. Like the seals that once climbed upon it, the lonely rock sits and gazes – a natural watcher of change abandoned from its former human purpose. I came here too, guided by the Åland Islands geodesist Martin Ekman. The author of a fine book and several scientific papers examining Celsius' geophysics, he had painstakingly identified and measured the rock from Rudman's 1731 drawings and Celsius' original dimensions. Without Martin I would never have found the forest path to this spot, let alone the rock itself. But in his company the scene came alive – I pictured Rik Nils, crouched in the tree line gripping his harpoon and waiting patiently for a seal to come into range.

It is easy to overlook that, as Celsius set about his task of cracking the Baltic mystery, much of the knowledge that might have helped him decipher exactly what was happening there was yet to be discovered. The layered composition of the planet, different climatic periods and glacial ice ages were then all 'unknown unknowns'. It would be another century before his scientific successors finally pieced together the complex interactions at work. And the truth proved to be far more intricate and amazing than even Celsius could have imagined.

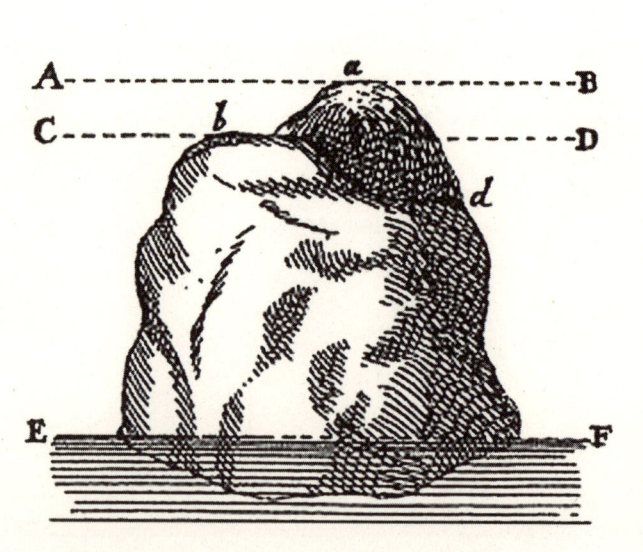

Celsius' drawing of the Iggön rock showing the fractures made by Rik Nils and changes in water level.

Celsius' first challenge was to work out what he meant, thought he meant or wanted to mean by 'sea level'. By definition, this could only be expressed in relation to something else. But should that be 'normal' sea level (an average based on lots of measurements over time) or a fixed point on the surrounding land at each spot? Both options presented difficulties for accurate measurement, since the comparisons could be variously affected by the water, the land or both. Ever the logician and empiricist, Celsius plumped for the simplest and most readily executed approach: what he could directly observe and record.

At Iggön, his calculations and interpretation of documents suggested an annual rate of mean sea level change of 1.4 centimetres. Even allowing for some mismeasurement or still hidden effect, this meant the sea level decreasing by around a metre every century, which roughly matched the accounts from Östhammar and Luleå.

While he mulled over his findings and what they signified, Celsius received further, relevant information from his long-time friend and university colleague, Carl Linnaeus, who was being assisted by Nils Gissler – a bright student who they had both tutored and mentored. Linnaeus and Gissler had also visited different spots on the Baltic shore and were intrigued by what they saw.

Three hundred kilometres south, on Gotland, Sweden's largest island halfway across the sea to modern Latvia, Linnaeus came across a sequence of ridges, running parallel to the coast. He deduced that these must be ancient beaches, which had long since lost contact with the sea. Meanwhile, further north in Härnösand, at about the north–south midpoint of the Gulf of Bothnia, Gissler was busy measuring not just the sea level, but also air pressure.

Gissler came from a modest farming family and had overcome great hardship to gain an education and forge an academic career. When he first came to Uppsala University, he had a future in the priesthood in mind, but he soon began studying an uncommonly broad range of subjects straddling logic, philosophy, mathematics, chemistry and – with Celsius, who had then just returned from the Arctic expedition – astronomy. No doubt influenced by the personalities and examples of his two professors, Gissler was a happy, generous, caring and cultured young man.

He too was fascinated by the sight of pebble beaches that now lay far from the sea. Might weather and atmospherics also have something to do with this? In his distinctive blue coat, tapering hat, buttoned knee-breeches and yellow stockings, he stood above the narrow sounds and bays around Härnösand, leaned on his thick staff and watched. In these sheer inlets, Gissler noticed that the waves sometimes seemed to pile up against the cliffs; then, as the wind dropped or changed direction, the water would rush out again.

From his observations, Gissler deduced an 'inverse barometer' effect. He estimated that (in modern units) for every increase of 100 pascals/1 millibar in local air pressure, the sea level dropped by a centimetre. He wrote: 'From the above observations I have found that, for the most part, whenever the barometer rises the sea level falls, and whenever it falls the sea level rises.'[3] Gissler speculated that the ridges he and Linnaeus had seen could have been created by severe storms, and then been slowly moved away from the sea by the gradual change in water level.

Spurred on by the observations and theories of his collaborators, Celsius wrote up his findings. Not content with measuring the rate of land uplift and/or water decrease, he asked himself what the longer-term effect of these changes would be. He prepared a huge, dizzying table on a single roll of the best paper he could afford. It charted total changes in sea level for 10,000 years into the past or future, assuming that the rate had been, or would remain, constant.

In his 1743 paper for the Swedish Royal Academy of Sciences, *Remarks on the reduction of water both in the Baltic Sea and the North Sea*,[4] Celsius wrote: 'If we knew here in Sweden the height above sea level of the most important places, … one would find that a long time ago … not only had Scandinavia been an island, but also that its southern part had consisted of small islands.' And contemplating the future, he went on:

> On the other hand, imagining what effect this lowering of sea level might have in the future, then the boundaries of Sweden would continuously expand. Our archipelagos would become gradually more filled with islands and rocks, so that pilots should measure the depth of the sailing routes at least every twentieth year, not trusting what their ancestors have made. In the long run, finally, the whole Baltic Sea would disappear.[5]

Celsius and his associates had accurately documented and described the Baltic Sea conundrum, and thought about the possible implications far into the future. But a complete explanation still eluded them. His paper conjectured about the causes of what they had seen:

The reason for this decline in water can be given in two ways. It is known that out of the sea a heap of fumes continuously rises, which then shuffle together to the mountains, and through large rivers also flow back into the sea. But it waters the earth and causes the plant of all the trees and herbs, so does not all come back to the sea, but remains in the plants. And then also with them, the solid or dry part of the Earth always increases, but the fluid, is continually diminished and at last would be utterly dehydrated, if not comets sometimes, with their close approach to our Earth, would, through their fumes, replace and fulfil it.

The second way of interpreting the evasion of the sea does not seem so unreasonable, if one imagines, that in the bottom of the sea there are one or more holes, through which the water gradually seeps down into the abyss of the earth.[6]

Fumes, trees, rivers, rain, comets, giant plug holes or an underworld abyss – despite all his meticulous examination and thought, Celsius was as far away from isolating the exact reasons for what was happening in and around the Baltic as most of his predecessors and contemporaries. But he had paved the way for an eventual solution and, thanks to a coincidence of timing and contacts, his opinions quickly attracted national attention.

Soon after Celsius presented his paper to the Royal Academy, the Swedish Parliament commissioned the poet, dramatist and historian Olof Dalin to write a history of Sweden. Six years younger than Celsius, from the Halland region on Sweden's south-west coast, Dalin had studied alongside Linnaeus at Lund University and was already an influential literary figure in aristocratic and royal society. Like Celsius, natural skill and intelligence backed by a strong academic lineage had helped Dalin to enjoy a Grand Tour through Germany and France. He had been elected to the Royal Swedish Academy of Science the previous year, and shared Celsius' free-thinking thirst for discovery.

Dalin seized upon his fellow academician's recent scientific findings and used them to frame the work that would become his four-part opus *Svea Rikes Historia* ('History of the Kingdom of Sweden').[7] The opening sentence read: 'The Nordic countries were still mainly under water, and were like an archipelago divided into a large number of small islands, when their highest areas were populated by the people, the history of which I am now going to write.'

Celsius was thrilled by this establishment exposure. Dalin's writing put him in the spotlight and brought his work to much wider notice than ever before. It even sparked intense parliamentary debate about what the ever-changing shape of Sweden would mean for its sovereignty, identity, defence, trade and diplomacy. But Dalin's references also brought criticism and opposition Celsius' way. The more pious parliamentarians were still hostile towards any idea that departed from scripture. They fulminated about how a large part of their country could possibly have been under the sea until a few thousand years ago.

In particular, the Lutheran bishop, physicist and botanist, Johannes Browallius, came out swinging. He dismissed Celsius' theories as little more than random variations or methodological errors. Born in 1705, and also a member of the Royal Academy since 1740, Browallius was another significant peer, who mounted a fierce biblical defence against Celsius' findings.

Browallius had formerly been a close friend of Linnaeus, two years his junior, who had named a genus of pretty, violet flowering plants, *Browallia*, in his honour. But things turned sour between the two men over their competing affections for the same woman. In 1737, Linnaeus was already engaged to Sara Lisa Moraea – a delicate beauty whose refinement also caught the bishop's eye. Sensing an opportunity to lure Sara Lisa his way, Browallius encouraged Linnaeus to travel and complete his studies abroad then 'marry some rich girl'.

Like Celsius in many ways, the writer and historian Olof Dalin (1708–63).

Johannes Browallius (1707–55) – chief critic of Celsius' findings about sea levels in the Baltic. Portrait by Margareta Capsia, 1750.

Linnaeus duly spent the winter of 1737–38 in the Netherlands, England and France – an absence that Browallius wasted no time in exploiting to build up a cosier relationship with Miss Moraea. While still abroad, Linnaeus received news that 'his best friend B' had taken advantage of his absence to woo his fiancée. By all accounts, Browallius had also done his utmost to persuade Sara Lisa that her betrothed would never return. But the scheme failed. Linnaeus came back to Uppsala and married Sara Lisa in 1739, his bond with the bishop forever ruptured. And in an act of taxonomical revenge, Linnaeus later added the pointed suffixes *demissa* (cessated) and *alienata* (stranger) to two individual species of the *Browallia* flower genus.

Celsius suspected that the bad blood between the bishop and Linnaeus was at least part of the reason for his very public and vehement critique. A decade before, a less world-wise Celsius might have been wounded by such confrontation, but now more experienced and secure in his standing, he shrugged it off. Science, he reasoned, would always trump faith with evidence.

Celsius left behind his marks, measurements and sea-level tables 'in order to make future generations able to determine this rate of [sea] change more accurately'.[8] His wish was fulfilled in the centuries that followed, as other scientists returned to the region and were eventually able to explain the interwoven causes behind the water's apparent disappearance.

The seal rock at Lövgrund today bears two further marks, made at century intervals in 1831 and 1931, each about 70 centimetres apart, suggesting a more or less constant and natural process. The water now lies ankle deep at the bottom of the rock – barely enough for a tricentennial measurement and carving in 2031: destined to be the last unless the effects of climate change-induced polar ice-melt begin reversing the trend back up the rock's craggy face.

〜〜〜

In 1774, thirty years after Celsius' death, more carving was under way. A dust-covered stonemason crouched on a wooden gantry lashed beneath the double-beamed lifting bridge spanning the sluice into Stockholm harbour. Beside him, fresh water from Lake Mälaren trickled through cracks in the heavy wooden lock gates, and above him the city's merchants and traders crossed the bridge, oblivious to the craftsman below. Onto the vertical face of the sluice wall he was creating the precise graduations of the Stockholm Sea Level Gauge, from which, by order of King Gustaf III, surveyors began collecting data later that year. The measurements have continued ever since, making the Stockholm Series the world's longest and most reliable record of sea levels.[9]

The spot where the mason chipped away at the wall was an apt choice. Stockholm was itself a consequence of the still-yet-to-be-explained shift between land and water. Lake Mälaren was a substantial body of water reaching 120 kilometres inland, flecked with hundreds of islands. But it had at one time been a coastal bay. By 1200 CE, its mouth was so shallow that ships had to unload their cargoes at the entrance. As the bay became a lake, the unloading point became a port and then the nation's capital. The sluiceway had been dug 100 or so years before the mason completed his work, making navigation possible again between the sea and Sweden's interior.

Later overtaken by automated, mechanical mareographs and then digital technology, the meticulous tables of the Stockholm Sea Level Series built on Celsius' methods and provided a further step towards understanding the Baltic enigma. The next breakthrough came from the accumulated work by a collection of scientists who postulated a much colder Earth at various points in the distant past – the ice ages.

The Danish-Norwegian geologist and mineralogist Jens Esmark was the first to make the mental leap between the glaciers still visible and grinding their way through mountain ranges across the globe and the notion that these must have covered far greater areas in the past. In 1798 he climbed Snøhetta Mountain in southern Norway – the first person to ascend its sloping, snow-capped cone. From the 2,000-metre summit, Esmark gazed out at the tell-tale signs of glaciation. The sun behind him lit up huge erratic boulders and the tortured creases of lateral and medial moraines winding into the valley below.

Identifying and explaining this landscape was a staggering insight in itself, but it led to further, far-reaching discoveries. A generation later, scientists realised that if the Earth had undergone such enormous climatic shifts, this might alter both the planet's tilt on its axis and the path of its orbit around the Sun. This in turn would affect the total amount of sunlight hitting Earth, and so continue driving natural climate variations in a continuous feedback loop.

Today we know these planetary wobbles and their impact upon the climate over tens of thousands of years as orbital forcing. The path from Esmark's mountain observations leads to modern scientists being able to predict the timing and likelihood of future ice ages with (by geological standards) reasonable precision. Others followed in his footsteps. The dashing Swiss naturalist and engineer Ignaz Venetz also concluded, in his 1821 work, *Mémoire sur les Variations de la température dans les Alpes de la Suisse*,[10] that much of Europe had at one or more times been buried beneath ice several kilometres thick.

A few years later, the German botanist Karl Friedrich Schimper took a whole-planet view of glaciation – noting identical physical features across Asia and the Americas. From this, he ventured that, as well as ice ages and temperate periods lasting a few tens of thousands of years, there must have been much longer hot and cold eras over millions or even billions of years. We now call these alternating, aeon-spanning intervals 'icehouse' and 'greenhouse' periods.

Schimper's younger brother Georg and his cousin Wilhelm were also gifted botanists, equally fascinated by the evidence of ancient times hidden in plants, fossils and rocks. They devoured the mounting evidence of ice ages, as explorers began unearthing almost intact remains of mammoths, sabre-toothed tigers and other exotic creatures from the Arctic permafrost.

Perhaps some sibling rivalry or inferiority made the balding and bespectacled older scientist reluctant to write up and promote his ideas, since he published little. But he did at least discuss them with his Swiss-born biologist contemporary, Louis Agassiz. In his 1840 *Études sur les glaciers*,[11] Agassiz presented his own '*Eiszeit*' findings, while consolidating and popularising the ideas of his predecessors to become the person most associated with the discovery of ice ages. After an intense century of science, another key piece of the puzzle fell into place.

～

While this cohort of scientists studied what was happening on the Earth's surface and the story it told, others looked deep below, into the planet's very core. The final, crucial element in explaining what Celsius had observed at Iggön and Lövgrund depended on knowing what the Earth was made of and how it behaved. Was it solid or liquid? Uniform or layered? Cooling or heating? Static or elastic? The answers, it turned out, were – over different time periods and at various depths – all of these.

In 1865, a few hundred kilometres west from Iggön and Lövgrund across the North Sea, the Scottish geologist Thomas Jamieson picked up the baton to consider what had happened after the glaciers melted at the end of the last ice age. He estimated this to have been around 12,000 years before. As a schoolboy, the Aberdeen-born and -educated Jamieson had exchanged letters with Charles Darwin and other star names of the time, including the Scottish geologist Charles Lyell (1797–1875), who had already returned to examine Celsius' seal rocks. Could these scientific heroes help settle the huge questions raging in Jamieson's inquisitive mind?

Fifty years later, on a hill high above the water in the Forth Valley leading to Scotland's capital Edinburgh, Jamieson picked through a handful of tiny marine fossils he had just collected at his feet. The presence of these organisms here convinced him about two things. First, this ground must have once been beneath the sea and, second, it was the disappearance of glaciers that had made the land rebound and spring upwards; a slow-motion ascent that was still in progress. Like Gissler and Linnaeus in Scandinavia before him, Jamieson went on to locate dozens of raised shorelines around Scotland, up to 100 feet above the sea. He had identified post-glacial uplift or, as it later became known, glacial isostatic adjustment.

Changes in land and sea levels do not just happen where the surface is relieved of the enormous weight of melted ice; what goes on underneath also matters. Modern seismology has shown that the Earth's inner core is solid: a 2,500-kilometre diameter ball of iron and nickel, squeezed hard by gravity and super-heated to over 5,000 degrees Celsius – about the same temperature as the surface of the Sun. The high temperature here comes partly from heat generated when the Earth first formed from post-Big Bang dust particles, gas and asteroids, and also the radioactive decay of uranium, thorium and other elements inside the planet.

Cloaking the inner core is another, similar thickness fluid layer of boiling metal, in which violent, churning waves generate the Earth's magnetic field. This then gradually gives way to the stiffer substance of

King's Caves on the Isle of Arran, Scotland – one of the raised beaches identified by Thomas Jamieson (1829–1913) – the Scottish geomorphologist who provided the final piece to the Baltic Sea puzzle.

the mantle and then up to the rigid, brittle crust on which all life exists. The cool, outer layer is just a few kilometres deep in places, and nowhere more than 60 kilometres. It is where unceasing plate tectonics shove and reshape continental landmasses at roughly the rate our fingernails grow. The Earth's crust is a thin veneer separating us from the molten world below, punctured intermittently by chains of volcanoes and sea vents – pyrogenic portals into the underworld imagined by ancient civilisations.

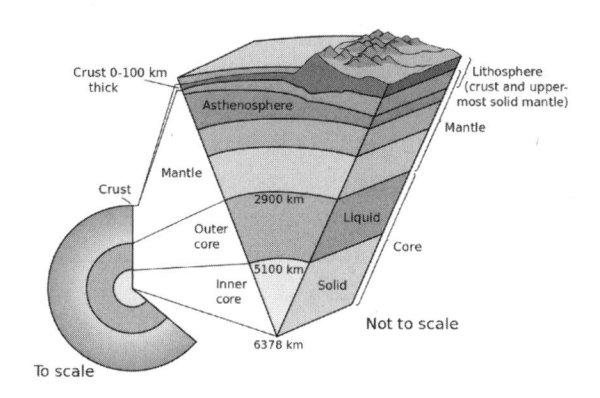

Our multilayered Earth. What goes on at the core affects everything else, up to the surface and out into the atmosphere.

The way these layers interact with what is going on above and out into the atmosphere and space drives plate movements, climate changes and glaciation, so eventually contributing to the phenomena of land uplift and sea-level changes. Rotation of an oblate planet means that the rate of spin at shallower, shorter latitudes nearer the poles is less than at the equator's maximum. This is due to the conservation of angular momentum – like a spinning figure skater drawing their arms in close to their body to make them turn faster. In a layered, semi-fluid Earth, the centrifugal force created by this 1,600 kilometres per hour revolution dictates the strength and direction of gravity and magnetism.

※

Twenty thousand years ago, the Lövgrund bay where the grey seal, Celsius, Lyell and I visited was covered by at least 2 kilometres of ice. This was the Weichselian ice age – the peak of Earth's last glacial maximum. As a great thaw began around 10,000 years later, the released water fed into the seas to drive weather patterns through evaporation. In the glacier-cut depression of the Baltic Sea, the planet's thin crust, crushed for millennia beneath its glacial load, began to rebound, set free from above and thrust from below by the viscosity of the mantle. This combination of forces gradually pushed the land further upwards to give the impression of falling sea levels.

This explanation – 900 years in the making from when Ingrid planted her rune stone at Runby – reveals the intertwined reasons behind the phenomenon that captivated Celsius. It is a wonderful illustration of his world-changing impact: a magnificent synthesis of his studies in geodesy, physics, mathematics, meteorology and astronomy, together with his collaborative approach and exceptional gifts for accurate measurement and analysis of longitudinal data. As Celsius' wide-of-the-mark initial ideas demonstrate, he was not afraid of being wrong about something – just perpetually excited and energised by seeking to uncover the truth. With many critical discoveries still far into the future, Celsius, his forerunners and most of his contemporaries viewed the Baltic question upside down. But the marks and theories he left behind made a telling contribution to reaching the eventual answer.

Depending on the counter-effects of human-made global warming, it is predicted that the next ice age will begin sometime in the next 15,000–100,000 years – most likely around the mid-point of that range.[12] Whenever the glaciers return, only a fraction of the human population and civilisation in existence at that time can be expected to survive. As our species has demonstrated many times before, it is supremely adaptable, but most of human life as we know it will disappear.

Against this background, people tend to make the same mistake as Celsius and his peers in the Baltic: looking at things the wrong way round. As the seas and temperatures rise because of our species' ever more numerous and rapacious presence on the planet, we talk about global climate change as the biggest problem facing us. But it is the opposite of that.

It is a vast truth, both profound and difficult. But *we*, not climate change, are the problem. Humanity has had its thumb firmly on the self-destruct button for over 200 years. And unless we make colossal, near-term changes we shall live – and die – with the consequences.

Our brains are ill-suited to comprehending natural processes that act over different scales and times – from invisible quantum forces to the still-expanding and accelerating universe, or nanoseconds to billions of years and light years. Yet we are the only species (as far as we are aware) that knows we will one day disappear. Whether that is because of the coming ice age or another natural process, catastrophic climate change, war, disease or a combination of these and anything else, matters little. We know dramatic change is coming.

18

THE INFINITE AND THE INVISIBLE

Magnetism and Making Connections (1740–43)

The visible world is the invisible organisation of energy.
  Heinz Pagels, *The Cosmic Code: Quantum Physics as the Language of Nature*, 1982

To investigate and better understand Celsius' science, I travelled to Amsterdam. A thirty-minute train ride west through serene countryside brought me to the coastal suburb of Castricum. A friendly, grey-haired man in a red coat met me at the station. It was Nicolàs de Hilster, an active member of the Scientific Instrument Society,[1] who had agreed to enlighten me about the surveying methods used by Celsius and his companions on their Arctic expedition, and to help me try out some of them first hand.

A short walk took us to the well-ordered home of Nicolàs and his wife Ria, who greeted me warmly and immediately produced some dark Dutch coffee and biscuits laden with fruit and spice. Sipping my coffee, I noticed a laptop on a side table. On its screen there was what appeared to be a monochrome image of the sun surrounded by tables of flickering data. Some sort of astronomical website I thought. But no, this was actually the live feed from Nicolàs' own observatory perched on top of the house, invisible from street level.

Coffee and biscuits finished, I followed my host up some spiral stairs to discover his amazing creation: a cluster of powerful telescopes mounted within a beautiful wooden dome – all hand-crafted from scratch by Nicolàs. Here was a modern-day equivalent of Magnus Celsius', Anders Spole's and Anders Celsius' home observatories in Uppsala. Nicolàs' instruments included a custom-made Galilean-type refractor, intentionally made to replicate the imperfections of seventeenth-century lenses.

Nicolàs pulled on some ropes, which caused the dome to rotate and a viewing window to drop with a satisfying clunk. Bright spring sunshine poured in and lit up the dome. Nicolàs jumped up onto a footstool and attached a shiny filter to one of the modern telescopes.

'Take a look through there', he said. I took his place on the stool and found I was looking directly at the midday sun – my eyes protected by the special filter. I saw a bright, fiery orange ball.

'Now adjust that knob and you'll be able to see some solar flares', said Nicolàs. And sure enough, as I turned the focus wheel, little looping swirls appeared around the edge of the image, clearly moving in real time.

'Wow!', I said, amazed and moved by what I was seeing. 'How big are they?'.

'Oh, about five times the size of Earth', replied Nicolàs.

I had mainly come to learn about terrestrial measurement, but Nicolàs had diverted me far away, to the cosmic realm that so engrossed Celsius and his ancestors. And I could understand why – it was magnificent.

The Sun and its effects upon the Earth preoccupied much of Celsius' time and thoughts once he moved into the Uppsala observatory. In particular, he and his assistants delved into the planet's magnetic field and how this connected with another longstanding fascination: the Northern Lights. While most of the other phenomena that Celsius had studied – from the Arctic landscape and Baltic Sea levels, to the influence of temperature and air pressure on weather patterns – were clear to see, here was something different. Magnetism was an invisible, all-permeating force of immense power that made organic life on the planet possible and safeguarded its survival.

As so often before, Celsius began his studies by working from familiar surroundings with devices designed and built by people he trusted. In 1740, in the garden of his mother's house, he set up a giant magnetic needle, which he had ordered from George Graham while he was in London four years before. The rectangular wooden case contained what looked like the central third of a compass – the sides cut away to leave a barrel-shaped middle portion of polished brass. Across this, suspended at the centre, hung the 30-centimetre-long needle, black, stiletto thin and pointing to the etched graduations at both tips.

Celsius was looking for declination – the difference between the current location of magnetic north (indicated by the trembling needle) and true rotational north: the axis of the planet, which he could deduce by

The Infinite and the Invisible

Daniel Ekström's copy of the outsized needle crafted by George Graham to measure Earth's magnetic fields.

observing the Sun or stars at their highest points. Anxious to avoid anything metal like keys, buttons or buckles that might skew the needle's reading, Celsius emptied his pockets, shed his coat and took off his shoes. He also moved the needle around the garden, noting the slightly different readings at various spots.

Back at the observatory, Celsius sat down to compile and analyse the results. By taking an average of all the readings, he concluded that magnetic north (and therefore the horizontal angle of the Earth's magnetic field) was 8° 49' west of true north, plus or minus 3'. This instantly caught his attention, because he remembered that a similar experiment carried out at Tornio during the Lapland trip had produced a figure of 5° 05'. The Earth's magnetic field evidently varied according to place and over time.

To cross-reference his readings in Gunilla's garden, Celsius took out another Graham instrument – a smaller brass dip circle, which resembled an upright compass needle mounted within a brass ring. This would give him an idea of inclination – the vertical angle of the Earth's magnetic field relative to the horizon. More measurements and more calculations yielded an average figure of 74° 05' – again different to what he had observed further north in Tornio, where the reading was 78° 05'. What could explain the difference?

In London, George Graham was conducting similar experiments. And he found something even more perplexing: that declination varied slightly when measured at exactly the same place on different days, or even at different times of the same day. On hearing about this from his friend in England, Celsius began to observe the needle hourly, wondering if weather conditions had some part to play. But no, he soon discounted this theory, recording in his notes: 'I have still not noticed that cold, heat, differing air pressure, wind etc have anything to do with this change.'

The hidden hand acting on the compass and circle must, he deduced, be something to do with the Sun. And to test this, he called upon the enthusiasm and resilience of Hiorter. They moved the Graham needle to Hiorter's room in the observatory and carefully positioned it to minimise any possible effects of iron in the doors and windows. For the next forty-six weeks, all through the late summer, autumn, winter and into spring, Hiorter kept up a punishing regime of twenty observations a day, often working by candlelight and snatching sleep when he could. He amassed 6,638 separate observations.

A brass dip circle similar to those used by Celsius and Graham to measure the vertical angle of the Earth's magnetic field relative to the horizon.

From these and his own experiments, Celsius pinpointed two other variables. Not only did the angle of declination depend on the time of day – varying by as much as 10' – it was also subject to much larger swings of up to 1°, apparently spontaneous and at random. In the first week of April 1741, the two scientists recorded some especially large changes in their readings. These prompted Celsius to write to Graham and ask if he too could observe a similar phenomenon, so checking that the magnetic needle itself was reliable. Graham replied quickly, and his records showed that he had witnessed exactly the same fluctuations in London. This meant that whatever was affecting the needle could not just be a local effect; it had to be something much bigger.

☾

When Celsius and Hiorter scrutinised their records, they realised there was a strong correlation between the biggest swings of the magnetic needle and the most spectacular displays of the Northern Lights. It was another six years, and three years after Celsius' death, before Hiorter finally published the revelation that drew these results together, and he generously attributed the discovery in equal measure to his late professor.[2]

In that week at the beginning of April 1741, Hiorter's diary records bright blue, green, purple and pink curtains of light, covering half the night sky and even appearing to the south. They could not yet pinpoint the force that created these displays, but they realised it was connected to the magnetic changes that caused the location of the North Pole to wander. Today we know these effects are generated by the constant solar wind and solar storms of charged particles streaming through space to collide with the Earth's magnetic field and light up the heavens.

This insight joined another set of dots in Celsius' work. It revealed that the same non-homogenous make-up of the Earth that contributed to its oblate form (as he had helped to prove in the Arctic) was also responsible for its strong magnetic field. Later generations of scientists would confirm that the planet's magnetic forces are generated deep within its liquid outer core, where a dynamo effect from the Earth's rotation stirs oceans of molten iron and nickel. Our home is, in effect, a massive magnet, projecting its arc far into the solar system, with radiating fields of invisible energy running pole to pole.

It is this that explains why the aurora borealis and aurora australis occur mainly at the polar extremes. Most of the electrons ejected from the Sun's superheated corona are deflected by the magnetic field, but at the poles, where the lines of force become concentrated, charged particles are directed so that they sneak through to excite the atoms in Earth's upper atmosphere. As the particles head to the poles they accelerate – reaching over 70 million kilometres per hour, slamming into other molecules that heat up very rapidly, enter their own high-energy orbits around their nuclei and then cool again. And it is as they cool that the gases glow – green for oxygen and pink, purple or blue for nitrogen. In 1619, Galileo named these awe-inspiring displays after the Roman goddess of dawn (Aurora) and the Greek goddesses of the north and south winds (Borea and Auster). The science of how they occur, hundreds of kilometres above the Earth's surface, is even more fantastic than mythology could suggest.

This beautiful phenomenon also occurs on other planets with atmospheres and magnetic fields – including Saturn, Jupiter, Neptune and Uranus. But in our solar system only Earth has the triple sweet spot of magnetism, make-up and mass that allows the Sun's deadly radiation to be kept at bay and organic life to flourish. This combines with our planet's 'Goldilocks

Zone' position, in which liquid water can exist on the surface. It is this special blend of make-up and circumstance that means there is a living and conscious audience for the auroral displays visible from our planet.

☾

While Celsius and Hiorter were immersed in their measurements of magnetic declination and inclination, another invisible agent was at liberty. The danger this time came from within: Celsius was infected with the then unknown *Mycobacterium tuberculosis* – the 'White Death' of consumption, which at the time was reaching epidemic proportions across much of western Europe. In contrast to the vast forces surrounding the Earth, the agents of this genus are microscopic. While a human hair is roughly 50μm (microns) in diameter, the average tuberculosis bacillus measures 2μm. But this tiny organism carries lethal force.[3]

At the peak of his intellectual powers and with the underlying connections between his previously disparate studies beginning to emerge, Celsius had very little time left.

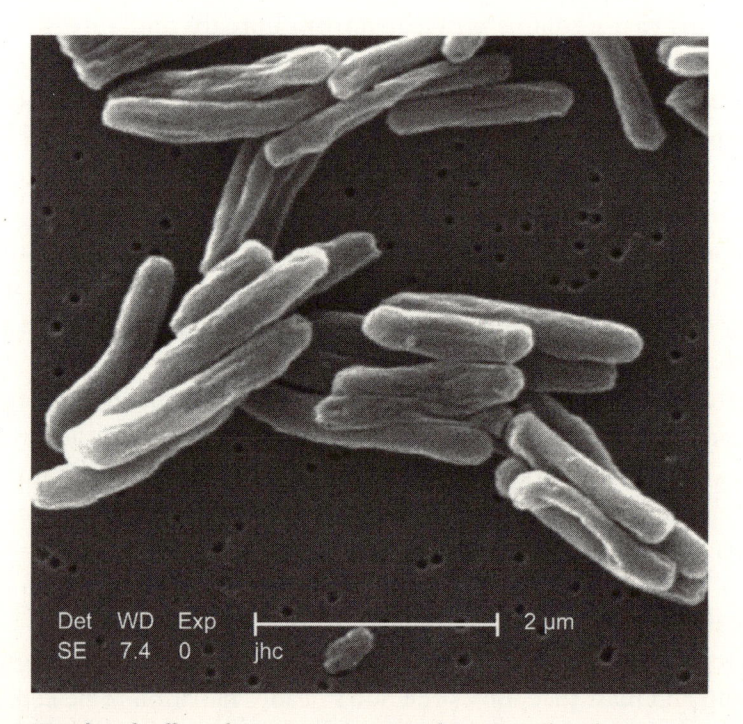

Tiny but deadly – the microscopic *Mycobacterium tuberculosis*.

PART V

TEMPERATURE AND CLIMATE

19

ONE HUNDRED STEPS

Creating the Centigrade Scale (1741–43)

Measure what is measurable, and make measurable what is not so.
Attrib. Galileo Galilei (1564–1642)

Christmas Day, 25 December 1741. If there is a single date that ensured Celsius' name would be known and remembered long after his death, it was this day, which saw the first recorded use of his eponymous 100-point temperature scale. Despite the holiday season, Celsius and Hiorter were busy on the second floor of the observatory in Uppsala. Five months earlier, they had taken delivery of the wall-mounted pendulum clock from George Graham in London.

Through the glass panels in the clock's walnut case, Celsius and Hiorter carefully logged the arc of the pendulum, noting the precise time it took from end to end on the multidial clockface above. A fire burned in the corner hearth, keeping the room's temperature as close as possible to what it had been when they began the experiments in the warmth of July. Even with Graham's expert design and construction, the slightest difference in the surrounding temperature would affect the density of air in the room, alter the length of the metal pendulum rod and skew their readings.

It was quiet, studious work, but for Celsius the steady swing of the brass weight amalgamated all his research and theories of the previous decade. The Graham clock enabled the scientists to measure gravity – the period of the pendulum's travel indicating the value of gravity in that particular place. Since the Arctic expedition had shown that the Earth was flattened at its poles, it followed that gravity here in Uppsala would be greater than nearer to the equator.

An entry in the Uppsala Weather Series from June 1743 recording the first outdoor use of the Celsius temperature scale. The 80.7°C reading corresponds to 19.3°C in the modern version.

Without fanfare or explanation, a new column and symbol appeared in their notes from that day. It marked the temperature inside the room according to the professor's own, brand-new scale: degrees Celsius. And eighteen months later, as part of the ongoing Uppsala Weather Series begun with Burman in 1722, Celsius applied his invention to record outdoor temperatures. Then began a zig-zag path spanning two centuries, to the 100-point scale becoming the near-universal global standard of measurement.

When Celsius turned his attention to questions of heat and cold, thermometry (from the Greek θερμός [*thermos*, hot] and μέτρον [*metron*, measure]) already had a rich history, dating back to the ancients. It was a scientific field filled with eccentric and exceptional characters, all following different paths to the same destination – how to quantify absolute temperature reliably. In the second and third centuries, the Greek/Roman philosopher, physician and surgeon Galen (129–216 CE) experimented with combining equal quantities of ice and boiling water to pinpoint a 'neutral' temperature. He saw heat as the vital fuel for animal life, forged in the act of conception and then gradually dissipating towards the cold of death.

At Padua University in Italy around 1593, the astronomer and physicist Galileo Galilei invented the thermoscope – a thin-necked vertical glass tube standing with its open base in liquid and an air-filled bulb at the top. As the ambient temperature rose, the air in the bulb expanded, pushing the liquid further down the tube. The thermoscope clearly showed changes or comparisons in temperature, but without any scale of graduations, not actual quantities.

In the mid-seventeenth century, Grand Duke Ferdinand II of Tuscany created the first sealed thermometer, filled with distilled alcohol inside which the movement of bubbles indicated changes in temperature. To this he attached a scale, divided into 360 'degrees' to match the emerging orthodoxy of geometric measurements. Others followed his example, all eager to address what became known as 'the thermometer problem': to create a device that could reliably quantify heat and cold. Many of Celsius' colleagues and other important figures whom he had met on his European travels applied their minds to the question. Klingenstierna, Kirch and Halley all worked on this elusive concept, as well as the Danish astronomer and Royal Mathematician, Ole Christensen Rømer (1644–1710).

Rømer is best known for making the first measurement of the speed of light in 1676. But before and after that he carved out a career of extraordinary variety. He was for a time employed by the French 'Sun King', Louis XIV as personal tutor to the Dauphin – the future Louis XV, who would later play such a vital role in de Maupertuis' and Celsius' Arctic exploits. Rømer's position at court led to him being involved in the design and construction of the magnificent fountains at Versailles. Then in a period of particular creativity around the turn of the seventeenth and eighteenth centuries, he introduced the first national system of weights and measures in his native country. He based this on the 'Danish foot', 24,000 of which made up a 'Danish mile' (equivalent to around 7.5 kilometres).

In 1699, Rømer was appointed Master of the Mint – a job that gave him a special interest in the melting points of the different metals used to manufacture coins. A year later, he persuaded the Danish King and state to switch to the Gregorian calendar and, a few months after that, restless while convalescing from a broken leg, he built what were effectively the first modern thermometers. He filled his prototypes with a mixture of alcohol and pure water or linseed oil, and graduated them with a scale running from 7.5 degrees (freezing) to 60 degrees (boiling). Between and beyond these points he marked eighteen other references, from human body temperature (12) to wax melting in water (24), the melting point of lead (90) and the heat of glowing iron (192).

Next up in the thermometry hall of fame was the Frenchman René Antoine Ferchault de Réaumur (1683–1757). He was principally an entomologist studying the diversity and behaviour of insects, but he was enthralled too by physics and mathematics. In 1730, Réaumur built on Rømer's experiments to produce ethanol-filled thermometers with an octogesimal (eighty-point) scale. Here, 0 degrees represented the liquid's freezing point and 80 degrees its boiling temperature, with the volume of expansion in the capillary tube indicating different temperatures.

Within its own terms, the Réaumur scale was a consistent and reliable measure, and it quickly became the most widely used – one that is still applied to some French and Italian cheese-making to this day. But variations in the quality and manufacture of early thermometers meant there was no guarantee that two instruments marked with the same scale would agree. In 1693, when grappling with these discrepancies, Sir Edmond Halley complained: 'Every thermometer works by Standards kept by each particular Workman, without any reference or agreement to one another … whatsoever Observations any curious Persons may make … cannot be understood, unless by those who have by them Thermometers of the same Make and Adjustment.'[1]

Sixteen years later, a Polish-born German physicist, living in the Dutch Republic, echoed Halley's frustration and resolved to do something about it. His name was Daniel Gabriel Fahrenheit.

Fahrenheit was born in 1686 into a successful merchant family living in Gdansk, now in Poland. His life took a tragic turn in 1701, the year Celsius was born, when his parents died after eating poisonous mushrooms. Fahrenheit continued his commercial training in Amsterdam, but became steadily less enamoured with money making and more fascinated by both natural sciences and the local craft of precision glass blowing.

Through travels to Berlin, Leipzig, Dresden, Copenhagen and in his hometown of Gdansk, Fahrenheit soaked up theory and practice, encountering some of the era's foremost philosophers such as Christian Wolff (1679–1754) and Gottfried Leibniz (1646–1716). He also met and became friends with Rømer and the young Swedish botanist, zoologist and physician from Uppsala, Carl Linnaeus. These experiences and encounters inspired a can-do sense of self-belief in Fahrenheit, and the last of them brought him to just one degree of separation from Anders Celsius.

Starting in 1709, Fahrenheit's early work used alcohol-filled thermometers. For zero on his scale, he took the temperature of a mixture with equal amounts of ice and salt. He then plotted and calibrated the melting point of ice (30 degrees) and normal human body temperature (90). Later adjustments revised these values to 32 and 98.6, with the freezing-to-boiling interval of water redefined to exactly 180 degrees, a highly divisible number convenient for calculations. Fahrenheit was also intrigued to notice that the melting and boiling points varied with changes in atmospheric pressure.

As far as is known, the two dominant personalities of modern thermometry, Fahrenheit and Celsius, never met or even directly corresponded. But their minds and methods occupied similar intellectual territory, and it is likely that, with their respective standing in European science, they were at least aware of each other. Quite independently, though, they gradually homed in on what it took to measure temperature accurately, and what this meant for our understanding of the world and universe.

When Celsius placed two Réaumur thermometers next to each other in the Uppsala University garden in 1732 to continue the measurement series begun a decade before, their readings differed by 12 degrees. Such inaccuracy and absence of reliability would never satisfy the perfectionist Celsius. He too decided to address the problem and, once he was back from the Arctic, he turned to another person he had never met, but who he thought could help him.

During his time in Paris, Celsius had stayed with the wife and daughter of the French astronomer and cartographer Joseph-Nicolas Delisle (1688–1768). Before Peter the Great had called Delisle eastward in 1725 to inaugurate the Saint Petersburg Observatory, the Frenchman had also developed a temperature scale in his own name. He took the boiling point of water as one fixed point (0 degrees Delisle), with the less precise minimum temperature of the Paris Observatory basement as the other. And when he met with the much colder conditions of Russia, he revised the lower range to equal the winter extreme of the basement at St Petersburg Observatory (150 degrees Delisle).

Apart from making larger numbers colder (and vice versa), Delisle adopted another practice that would soon influence Celsius' thinking and solution. Following the experiments of Florentine makers as far back as the 1650s, he tried filling his thermometers with mercury.

Celsius was familiar with the unique properties of this strange substance – the only metal that is liquid at room temperature. If he placed a droplet of mercury onto a flat surface, it instantly formed a raised dome – not collapsing and spreading like any other liquid or staying solid and still like any other metal, but a dense, vivacious bubble, sitting proud and ready

Some of Celsius' precursor temperature theorists and thermometrists (from top, left to right, Galen, Galileo, Ferdinand II, Réaumur, Rømer, Fahrenheit).

to respond to the slightest change or stimulus. He was also fascinated by mercury's shape-shifting sheen – one moment the shiny white of polar ice, the next, a shimmering graphite grey or mirrored silver like polished chrome, constantly reflecting and upending everything around it.

Celsius also knew from experiments by others that when mercury came into contact with gold, silver, copper or tin, something even more remarkable occurred. Without heat, pressure or any other agent, it would dissolve the chemical bonds of the adjacent material, fusing itself and the other metal into a new alloy. Could this quixotic but toxic metal be the answer to the final part of the thermometer problem?

To put things to the test, in 1737 Delisle made and sent to Uppsala two mercury thermometers marked with his own scale. He invited Celsius to compare their accuracy and how this related to measures achieved with other scales and instruments. But when the consignment arrived, one thermometer was broken, the mercury from its cracked bulb running loose around the straw in which its slender case was packed.

Luckily the second was intact – a simple rectangle of coarse wood, 30 x 5 centimetres, with an elegant pear-shaped reservoir of mercury and a slim tube fixed to the base by twists of jute twine. This delicate instrument was Delisle's when it arrived in Uppsala, but Celsius was about to make it his own. And it would alter the way humans perceive and respond to their environment forever.

A fierce 'quantifying spirit' characterised this period of the Enlightenment.[2] But even in this context, Celsius' decision to take the melting point of ice *and* the boiling point of water as the benchmarks of his temperature scale was a seminal act of metrology. Like Fahrenheit, he knew from his and others' earlier experiments on air pressure that these points were not completely fixed, but varied according to altitude. So while basing his scale on conditions at one standard atmosphere (roughly the average global atmospheric pressure at sea level), Celsius was able to provide tables and formulae to allow for this. He calculated that for every 152.4 metres of altitude, the boiling point of water dropped by half a degree on his new scale.

Relating temperature to everyday phenomena both familiar and visible to everyone created an intuitive and accessible solution. We take it for granted now that water and other materials behave in particular ways at specific temperatures, but establishing this before there were trustworthy instruments to prove it took a giant leap of imagination.[3] Though even this breakthrough required refinement — work that others would continue long after Celsius' death. When exactly can water be said to boil — at the first sign of rising bubbles, or when its surface is engulfed in continuous rolling motion, or somewhere in between? And to what kind of water should these values relate — fresh water, sea water, pure, filtered or distilled?

Today, the original Delisle/Celsius thermometer is one of the most popular exhibits at the Gustavianum Museum in Uppsala — from the rooftop of which Olof Rudbeck the Elder had tried to direct operations during the city's Great Fire of 1702. The museum was closed for renovation when I visited, but I was able to arrange a private viewing. Curator Anna-Zara Lindbom lifted the precious object out of its hardwood case and laid it on the table in front of me. It is humbling to handle something so fragile and significant; it was several minutes before I had the confidence to pick it up in my blue-latex-gloved hands.

The brown paper strip gummed beneath the tube bears Celsius' distinctive handwriting. Just below the twist of wire threaded through two holes at the top, are the words '*Chaleur d'eau bouillante*' (temperature of boiling water) alongside the figure 0. One hundred degrees lies about two-thirds of the way down the tube, closer to the shiny mercury bulb. In developing his scale, Celsius had followed Delisle's approach of having 100 degrees as the melting point of ice and 0 degrees as boiling water — guided perhaps by the prospect of taking readings in a country where the temperature regularly fell below freezing, to avoid the extra complexity of working in negative numbers.

Exactly who was responsible for turning Celsius' invention upside down, and their reasons for doing so, are the subjects of some disagreement and debate. The inversion is variously ascribed to the French physicist Jean-Pierre Christin or Linnaeus, and also Wargentin or even Daniel Ekström.[4] But it is certain that the change to what we know and use today took place not long after Celsius' death, around 1747. In many respects, it does not matter — as a 100-point (*centi-grade*) scale, the concept works equally well in either direction, its genius born of its simplicity and consistency. And once again, Celsius' timing was fortunate; in 1739, Sweden introduced a national decimal system for both distance and (in part) its currency. The era of 'base ten' had dawned.

To share news of his temperature scale invention beyond the observatory and Uppsala, in mid-1742 Celsius submitted a paper to the Royal Swedish Academy of Sciences.[5] Excited by its potential, in a covering letter he wrote: 'One has reasons to believe that here at the [Uppsala] Observatory a meteorological journal will be kept ever after.'

Thus far, his confidence has been vindicated. Data from the Uppsala Weather Series continues in an unbroken line and, in the meantime, Celsius has become the official unit of temperature measurement in all but a handful of countries.

By 1750, the Celsius scale, more commonly known at the time as 'the Swedish thermometer', was in widespread use throughout Europe. Nineteenth-century scientists such as William John Macquorn Rankine (1820–72) and William Thomson (1st Baron Kelvin, 1824–1907) developed Celsius' work to create absolute scales of thermodynamic temperature, applicable not just to conditions on Earth, but anything and anywhere in the universe. Thus, one Kelvin is the same as 1 degree Celsius, but on a much bigger scale descending to the notional absolute zero of 0K (−273.15°C).

At this extreme end of measurement, things become highly abstract. Modern science suggests that the absolute zero point can never be achieved, because by the laws of atomic motion it would require an infinite

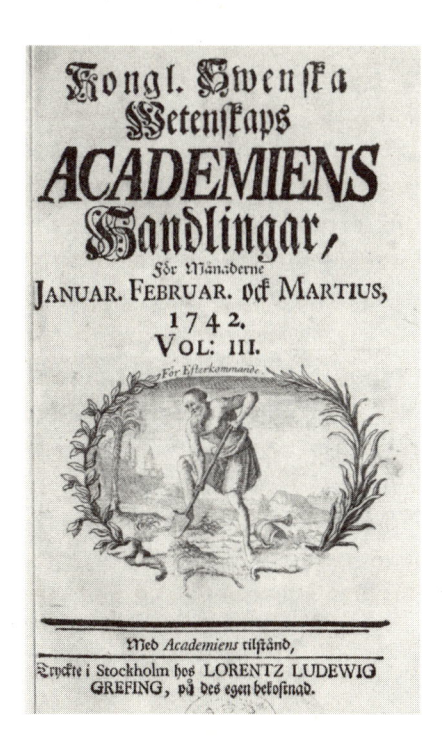

Title page of the Royal Society's 1742 digest in which Celsius announced the birth of his new temperature scale.

William Thomson (1st Baron Kelvin, 1824–1907) aged 22 – inventor of the SI unit of absolute thermodynamic temperature, used in engineering and physical sciences. One Kelvin is equal to one degree Celsius.

amount of energy to remove all the heat necessary to reach the absolute. But thanks to the theory and experiments involved in determining this, we now have particle accelerators to study the quantum building blocks of nature, magnetic resonance imaging scanners that can detect illnesses deep inside the body and near frictionless, superconducting materials with the potential to revolutionise computing, travel and many other aspects of life. Theoretical or otherwise, the science to which Celsius and Hiorter made their critical contribution on Christmas Day 1741 continues to have enormous practical impact.

On a thickly wooded hillside overlooking a tight bend in the River Seine, just west of Paris, stands the white, three-tiered Pavillon de Breteuil. It lies roughly halfway between the city centre and Versailles, from where, in 1785, Louis XVI's Queen, Marie Antoinette, acquired the palace and surrounding parkland as part of the royal estate. One hundred and ten years later, after the horrors and upheaval of the French Revolution had subsided, the pavilion became home to the International Bureau of Weights and Measures (BIPM).[6] And as a symbol of their lofty purpose, the building and gardens were formally designated as international territory. In the temperature-controlled basement there, visitors can still see the prototype platinum and iridium cylinder that defined the kilogram and a rod fashioned from the same alloy to delineate the metre. And it is here that the Celsius scale finally ascended to global dominance.

The ninth Conférence Générales des Poids et Mesures held at the pavilion in summer 1948 had a weighty agenda. The gathering was BIPM's supreme authority, originally formed from just twenty nations in 1875 and eventually recognised as the supreme adjudicator of global standard units of measurement (SI). As well as temperature, the 1948 delegates were called upon to define, decide on and classify SI units for electrical current (Ampere), charge (Coulomb), inductance (Henry), resistance (Ohm), potential (Volt) and capacitance (Farad), plus energy (Joule), force (Newton), power (Watt) and pressure (Bar). The decisions in front of the representatives reflected both the primal forces of nature, and the work of many of the most prominent and radical thinkers from the preceding three centuries.

Celsius' scale took its place among this eminent line-up. The conference's task was to avoid any confusion with the term 'centigrade', referring to both temperature and a division of the Gradian system of angular measurement (one-hundredth of a right angle) in French and Spanish. The solution, they concluded, was to follow the eponymic path and name the SI unit after its inventor. So degrees centigrade officially became, and remain, Celsius, °C. Like its inventor though, the Celsius scale's time as the sole international unit was short-lived. In 1954, it was supplemented for thermodynamic temperatures by Kelvin.

For some time after the 1948 Paris Conference, large parts of the English-speaking world held onto the imperial system, with Fahrenheit as their preferred unit of temperature. But they too gradually switched to Celsius for everyday use: India in 1954, the United Kingdom in 1961 and Australia in 1969. Today, only five countries retain Fahrenheit as their official standard: the United States, Belize, Liberia, the Bahamas and the Cayman Islands.

It may seem odd that the first and largest of these could be both a world superpower and an outlier when it comes to measuring temperature. The US Congress did pass a Metrication Act in 1975 to begin the process of conversion, but made it voluntary – and so ensured that most of the country's public, industries and institutions simply stuck to the familiar and ignored the change. For now at least, temperatures in the United States continue to reflect the country's colonial and Anglophone past.

Our planet of course takes no heed of how we measure the changes now being wrought upon it. The impacts of global warming and a degraded environment are the same however we choose to calibrate them. What is evident, though, is that the choices humanity makes in the period leading up to the centenary of the Celsius scale's formal adoption in 2048 will determine its future. How will history judge our progress? And what will we leave behind? These stories will be told in degrees Celsius.

The Pavillon de Breteuil outside Paris – where degrees centigrade officially became Celsius in 1948.

'This place is too cold in winter and very hot in summer', said Jörgen Runeby, copywriter at the communications company To Be Frank,[7] which now occupies Celsius' former rooms on the top floor of the yellow observatory building at 9 Svartbäcksgatan in Uppsala. 'Look how thick these walls are', he went on, pointing out the metre-deep, angled opening at each window. Jörgen and his colleague Petra Stärkman had invited me to visit their office and see for myself the place where Celsius devised his temperature scale. I was pleased to see a copy of Johan Henrik Scheffel's portrait of the 33-year-old professor hanging on the wall. The firm's black Labrador, Dylan, padded over to me carrying a blanket in his mouth, hopeful of some attention.

I had already visited the jewellery firm that leases the ground floor. There were no complaints there about the solidity of the walls: good security for the precious contents inside. But the friendly co-owner, Lena Urtel, agreed about the heat and cold – it is not the most comfortable place to work.

On both floors, we looked at the layouts together – comparing the modern arrangement and use of rooms with Hårleman's drawings. A wall has been moved here and a fireplace blocked up there, but the shape and features of the interior were mostly intact. The staircase looked much as it must have done in the eighteenth century – a gentle stone spiral with a sturdy metal handrail. I tried to imagine the massive hardwood document cupboard belonging to Celsius, which I had seen at the University's Ångström centre, being manhandled up to this second level.

The roof, alas, is no longer accessible, and looking up at the tiny hatch on the uppermost landing, it must have been a precarious climb even when Celsius, Hiorter and Wargentin were here. The building's manager, Karl Lagergren, sent me some recent photographs of the top of the building. There is no sign any longer of the six-sided viewing tower that once stood here, but there are two symbols of the modern era and its changing relationship with temperature: an air conditioning unit and an air source heat pump.

In one of the photos, just visible over the roof's parapet, the statue of Anders Celsius looks on from the street below – its fountain now restored and once again gushing water over the globe.

20

DEATH OF A STAR

Celsius' Illness and Death (1743–44)

For what is it to die but to stand naked in the wind and to melt into the sun? And when the earth shall claim your limbs, then shall you truly dance.

Kahlil Gibran, 'On Death' from The Prophet, 1923

After tens of thousands of years in the farthest reaches of the solar system, a town-sized mass of frozen rock, dust and gas tore towards the Sun. As the cosmic snowball's elliptical orbit reached its closest point to Earth, observers looked up and recognised the tell-tale fuzzy form of a comet. The first sighting of this particular object was over the Netherlands on 29 November 1743. Then astronomers all across Europe began to see it.

In Uppsala, Hiorter was the first to spot the comet. The next night, Celsius joined him on the observatory roof and saw it too. By Christmas night, the celestial visitor had spouted a luminous tail, and within a month it was the brightest thing in the sky, creating a mighty six-forked fan of light, which they measured as 18° wide. Always keen to attract political support for the University and his work, Celsius sensed an opportunity; he invited the heir to the Swedish throne, Crown Prince Adolf Frederick, to Uppsala to witness this spectacular event.

On 29 January, the observatory team put on a show for the royal visit. They decorated the outside of the building with lights to welcome the Prince and polished the best telescope for his use. The King-in-waiting was a gentle but weak and ineffectual young man, apparently more interested in his favourite pastime of making and collecting decorative snuffboxes than the marvels of the Earth, its solar system, galaxy and

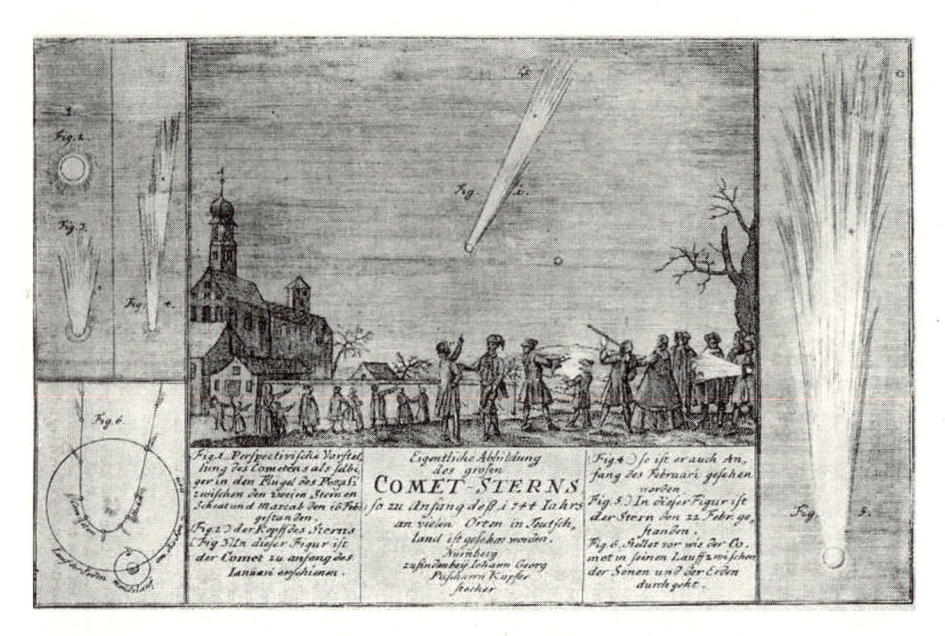

The flared tail of the Great Comet as seen over Nuremberg, Germany on 16 February 1744. Engraving by Johann Georg Puschner (1680–1749).

the stars beyond. But Celsius, ever courteous and mindful of his guest's rank and influence, escorted the silk-dressed Adolf Frederick up the tight staircase to the rooftop. There, he showed the Crown Prince how to look through the eyepiece of the largest tube and observe the comet's form and radiance at closer quarters.

A few months before, Celsius had been honoured (or, as he saw it, encumbered) with the position of university rector. There were painful memories of his father Nils' disastrous struggles with religion, but Celsius resented the appointment most because of its potential to distract him from his work. In a letter to Benzelius – who was by now the Archbishop of Sweden – he explained that he could not allow this new and unwanted role to affect his science. 'Rather they may complain that I am a neglectful rector', he added.

There's no record of what the Crown Prince thought about his visit and seeing the Great Comet of 1744, or how successful Celsius was in minimising his clerical duties. But as the ice ball swung away from the Sun and quickly faded from view, Celsius' health and strength also began to desert him. Less than four weeks after Adolf Frederick had viewed the comet at its most brilliant, Celsius peered into the same telescope for the last time, and made his last written entry. It was 24 February 1744.

Until the 1730s, it had been widely believed that tuberculosis could be healed by 'royal touch' – a sovereign with divine power simply placing their hand upon those afflicted to cure the disease. If so, Celsius' recent brief encounter with Swedish royalty had not helped him. His physical condition – never robust after the cold and hardships of the Arctic expedition – started to decline rapidly.

Gunilla, Sara-Märta and others did their best with the primitive treatments available, including frequent doses of cod liver oil, vinegar massages and inhalations of turpentine, but they had no effect. Death from tuberculosis at this time was drawn out, messy and painful. Consumption – the name by which the disease was popularly known – was an apt description, since once the infection took hold in the sufferer's lungs, the other internal organs would be steadily consumed, until the patient eventually drowned in their own arterial blood and body fluids. The 1745 memorial tribute to Celsius by the Hat Party Baron Anders Johan von Höpken[1] describes him suffering from 'a tearing fever and incurable pleurisy', as the tubercules multiplied and attacked his respiratory tract.

The Great Comet was last seen on 22 April 1744 and, three days afterwards, a clergyman came to visit Celsius on his deathbed at Gunilla's home. To prepare the patient for his end, the priest sat alongside him and talked in a soothing voice about the immortality of human spirit and the promise of a life after death. Pale and exhausted, Celsius listened, his eyes closed. Then he replied: 'Is that what you think, Sir? Well, quite soon I will be in a state where I will see whether that is true or not."

Celsius died at 7 a.m. on 25 April 1744. He was 42. As was once widely believed, the comet had seemed to be a harbinger of death.

One week later, a sombre gathering of family, friends and colleagues bore witness in the gloom as Celsius' body joined those of his father and grandfather in the family vault at the little church at Gamla Uppsala. Some of his astronomy students penned a poignant poem in their professor's memory, which captured the sense of loss among Celsius' family, the University and science itself. It began:

What's this?
Is evening now arriving,
When sun from highest is still shining?

Most of Celsius' pension from Louis XV had gone towards supporting the observatory, increasing his library and employing assistants. Shortly after her son's death, Gunilla made an inventory of his belongings. It's a meagre list, which records Celsius' only personal assets as:

One black gown and one brown gown (both turned inside out because of wear)
Two old red coats
Eight shirts
Three pairs of stockings
Three pairs of shoes
One pair of boots
A closed horse carriage (the gift of de Maupertuis after they returned from Lapland)
Assorted items of silver and wooden furniture.

But, most importantly, there were also around 1,500 books – the remnants of his father's and grandfather Spole's collections rescued from the Great Fire in 1702, plus the volumes that Celsius had acquired during his own lifetime. Gunilla decided that her son-in-law, Hiorter, should inherit the library. He, in turn, donated it to the University, where it remains protected by the security of the Ångströmlaboratoriet.

Mårten Strömer (1707–70) Celsius' somewhat unpopular successor as Uppsala's professor of astronomy. (Image courtesy of Gustavianum Art Collection)

It is impossible to know what Celsius and his colleagues might have achieved together at the observatory in Uppsala if he had he lived longer. But after his death, leadership and utility of the yellow building soon descended into acrimonious farce.

In the close competition to replace Celsius as Uppsala's professor of astronomy, Hiorter came third behind the mathematicians Samuel Klingenstierna and – the Senate's and King's eventual choice – Mårten Strömer. At 37, Strömer was eleven years younger than Hiorter, a student of Klingenstierna who had not even formally applied for the position. It was a disappointment and perceived snub that Hiorter found hard to shake off. He objected to Strömer's comparative 'lack of theory', and appealed to the University's Chancellor Gyllenborg that *he* should be allowed to lead the observatory in a new capacity of royal director, with the new professor not allowed to enter or be involved in its work. As an inducement, he even dangled the prospect of a substantial donation from his wealthy former student, Oxenstierna.

Staking his claim with angry sarcasm, in January 1745 Hiorter wrote a 'Humble Memorandum' to the Senate, objecting to Strömer's presence, habits and motivations:

> To occupy and inhabit a-compared-to-other-buildings distinguishing observatory, to drive and use a skilled helper as a lowly subaltern, to show up during observations as he sees fit and when least needed, but for the more fatiguing chores invoke academic duties, to shine in the eyes of the world with the help of the discoveries and the ardent work of the subaltern, and more such perhaps that should be seen as a distinction.[2]

In contrast, Hiorter set out the character and competencies that he believed the experimental astronomy going on at the observatory demanded: 'innate inclination, attention, good practice, diligence, carefulness and vigour in many fatigues also in the harshest season' and 'a peculiar nature and innate inclination, an unusual attention, through extensive and good practice acquired habitude, untiring watchfulness etc.' What the observatory did *not* need, he asserted, was 'a persona principali that has never shown the smallest sign of the above'.[3]

After so much bile, an uneasy compromise was eventually worked out. The Senate permitted Hiorter to live at the observatory and oversee its work with minimal interference from Strömer. Despite his lack of formal qualifications, Hiorter was also made an assistant professor and awarded the ego- and face-saving title of Astronomer Royal.

But, meanwhile, more practical concerns emerged. Despite its tall windows on every corner and every side, the observatory's central location meant that the view to the south was partially obscured by the cathedral and other buildings, rendering it useless for certain astronomical observations. The delicate instruments inside were also constantly shaken and jolted out of position by passing carts and carriages. And as the city grew, with its expanding population burning more fuel, the sky became progressively thicker with smoke, soot and dust. To overcome these problems, the astronomers began to use the more elevated and sturdier Uppsala Castle tower for their observations.

It seemed that, without the observatory's originator at its helm, the belief and investment the University had put behind Celsius' ideas were no longer paying back. By January 1750, even Hiorter was losing faith. He complained to Wargentin about a much-anticipated observation: 'I have not seen the comet in question, thanks to our beautiful observatory.' The following March, Professor Strömer wrote to Gyllengborg, concluding, 'observations must be made in some other place'.

That other place was Stockholm. In 1744, Pehr Elvius the Younger took over as secretary of the Swedish Royal Academy of Sciences and managed to secure something that had eluded Celsius: royal privilege to issue almanacs. This monopoly put Stockholm on a par with the Kirch family in Berlin, and established a reliable income stream that soon provided enough money to build a new, purpose-built observatory. Wargentin championed its construction and, in 1753, became its founding director. The new building, bigger, brighter and better positioned on top of a hill, was designed by the same architect, Carl Hårleman, who died shortly before its opening.

It would be another century before Uppsala had its own observatory again. And with the reflected light of Celsius' star waning, the yellow house on Svartbäcksgatan went through a range of different uses and occupiers as diverse as the night sky objects that it used to observe. At various times after Wargentin moved on, taking the focus of Sweden's astronomical research with him, it served as a chemistry laboratory, student hall, church, community centre, brewery, bottled water factory and choir headquarters.

The observatory building remains, like its founder whose modern statue-cum-fountain commands the pedestrianised streetscape, striking, out of the ordinary and, above all, important. Meanwhile, millions of kilometres away, the Great Comet – now less evocatively classified C/1743 X1 – continues its orbit. Measured both by absolute magnitude (0.5) and apparent magnitude (−7) it is one of the brightest celestial objects ever seen from Earth. Predictions vary, but if it does return again, it will likely be in many thousands of years' time. Who, if anyone, will be here to see it then?

21

NOBLE SUCCESSORS

How Wargentin, Hiorter, Strömer and
Others Continued Celsius' Work

To open the doors of truth and to fortify the habit of testing everything by reason are the most effectual manacles we can rivet on the hands of our successors, to prevent their manacling the people with their own consent.

Thomas Jefferson, letter to Judge John Tyler, 28 June 1804

As a master of geometry, calculus and statistics Celsius understood the importance of constants – those naturally occurring, enigmatic relationships that explain a host of realities. Without Archimedes' pi[1] (or Euclid's golden ratio (ϕ),[2] Celsius could never have made his crucial contributions to Enlightenment discoveries about temperature, magnetism or the shape of the Earth. And the mathematical relationships later named after Amedeo Avogadro (1776–1856) and Ludwig Boltzmann (1844–1906) also featured in his observations and calculations of air pressure at Sala, Falun and Uppsala Cathedral. But the most important constant in Celsius' life took human form: his mother Gunilla.

It was she who swept up her 6-month-old son to save him from the Great Fire of Uppsala, and she who helped to rescue her father's precious books to fuel Celsius' early fascinations and learning. And it was she who scraped together the means to open and run the eating house that paid for his university education. From the portrait of Gunilla painted around 1740, it is clear that Celsius also took after his mother in his looks. They share the same arched eyebrows, sparkling eyes, full lips and firm chin, and also a similar poise. In Johan Henrik Scheffel's painting, Gunilla's face is framed by a delicately folded lace bonnet, which spills over her shoulders to meet the emerald dress draped halfway across her torso. Slim shoulders relaxed, her right hand rests contentedly on something (or someone) unseen – one can almost imagine it being the arm of her beloved son.

Parents do not get to choose or dictate their children's characters and capabilities. But just as Gunilla concocted hearty, delicious meals for the city's academics, she and her husband Nils created the perfectly balanced recipe for a world-changing scientist. Abundant innate intelligence mixed with physical bravery, intellectual boldness and relaxed self-assurance brought Celsius to the brink of greatness. His skill at collaboration, boundless confidence, empirical flair, affable manner and comfortable internationalism secured it.

Gunilla added a final, all-important ingredient: the desire and drive to see her son succeed, free from the frustrations and setbacks that had so afflicted Nils' academic career. Gunilla's belief underpinned Celsius' determination. Her gentle manner fashioned his ease. Her love was his constant and closest companion.

Throughout Celsius' short life and the years in which he became a scientist of international repute, Gunilla's influence and the qualities she

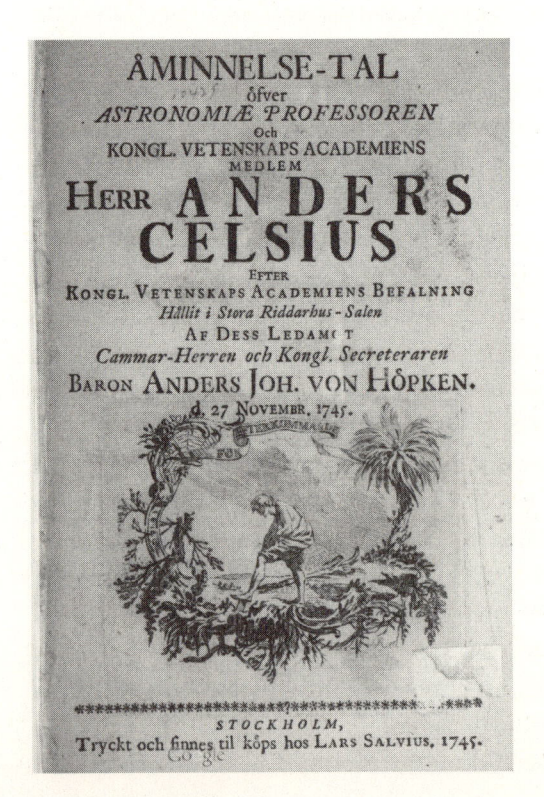

Cover of the 1745 memorial address to Celsius by Baron Anders Johan Von Höpken.

instilled in him were widely felt. No one who knew and worked with Celsius – from his student friend and later fellow professor Klingenstierna to his early mentor Burman, his Grand Tour companions Meldercreutz and Biurman, the dashing de Maupertuis and his expedition team, librarian-turned-archbishop Benzelius or faithful instrument makers Graham and Ekström – could ignore his winning mix of dignified charm, piercing intellect and never-ending curiosity.

In the last five years of his life, Celsius had a particular impact on the fortunes of those who worked alongside him. Hiorter did not live much beyond Celsius, but in the six years leading up to his own death, aged 54 in 1750, he became the spiritual, material and, to a large extent, academic heir to his brother-in-law. Watching the two men's relationship develop within the tight observatory team, Wargentin described their bond as harmonious and blissful:

> The close companionship of Celsius and Hiorter was founded on equal inclinations and ambitions and was strengthened through their being brothers-in-law so that one was aware of but one Man at the new observatory in Uppsala, which through their common efforts rapidly achieved a great reputation and soon seemed to want to compete with its older foreign brethren: because a Celsius and a Hiorter were not to be found together everywhere, though both are needed for important work to be done. But they were all too soon separated, while the expected fruit was still germinating.[3]

Hiorter had been Celsius' faithful and hardworking companion – someone willing to undertake the labour of countless observations and mathematics, and with the same single-minded vision to create a space for practical astronomy and professional science at the observatory. In a clever move, when donating Celsius' library to the University in 1747, Hiorter added a powerful condition to the deed of gift, 'that I may, until my death, be assured those liberties that I have been granted, personally and as to my possessions, for the benefit of the observatory and for science, and that I may under no circumstances be evicted [from the observatory] by no one, for any reason whatsoever'.[4]

While Hiorter shored up his position and sought to continue Celsius' work in Uppsala, Wargentin's fast-maturing statistical brilliance set him on

Bigger, better and purpose built; the new Stockholm Observatory opened in 1753 to overcome many of the problems inherent in its Uppsala predecessor.

a separate path to status and success. In 1746, Uppsala University awarded him an associate professorship and three years later, following the death of Pehr Elvius, Wargentin took over as secretary to the Royal Swedish Academy of Sciences – a position he was to hold for the next thirty-seven years. This role came with the added lure of establishing the new Stockholm Observatory, which was already under construction. A week after his appointment there, Wargentin wrote to his former colleague and tutor Hiorter, thanking him for 'every happy moment we have had together in Uppsala', and sending his fond wishes to 'the wives' (Gunilla and Sara-Märta).

Once the Stockholm Observatory opened in 1753, Wargentin embarked on a twenty-five-year obsession with studying the moons of Jupiter as an aid to determining longitude. He continued what he had begun with Celsius in Uppsala, but took his methods to new heights of accuracy and analysis. And, by calling on the contacts forged during Celsius' Grand Tour, Wargentin was able to compare his own observations of Jupiter's four largest moons – Io, Europa, Ganymede and Callisto – with those from Paris and Greenwich.

It was Galileo who had paved the way for this branch of astronomy. When he identified the four moons in 1610, they were the first objects shown to be in orbit around anything other than the Sun or Earth. Galileo noticed that the moons' apparently irregular eclipses and movements around the giant planet were actually calculable and predictable, providing a trusty celestial clock to pinpoint an observer's position on Earth. Following this principle, Wargentin calculated the difference in longitude between the observatories in Stockholm and Paris as 15° 43.2'. Modern satellite measurements show it to be just one-tenth of an arc minute greater – equivalent to about 200 metres on the ground, calculated from objects almost a billion kilometres away.

Wargentin achieved almost the same precision in calculating the longitude difference between Stockholm and Greenwich, following which astronomers from l'Académie des Sciences asked him to calculate the comparison between Paris and Greenwich. Without leaving his home country, Wargentin lit up the path Celsius had taken almost two decades before, connecting the points his late professor had plotted and reaping the benefits of his networks across Europe.

Next, Wargentin moved onto studying latitude – helping to realise Celsius' idea of organising simultaneous observations on the same meridian in the northern and southern hemispheres (in Stockholm and Cape Town, South Africa). These results, in turn, made it possible to calculate the distance from Earth to the Sun and its other planets. It was a moment of international scientific cooperation and astronomical achievement, which would surely have made Celsius proud. But Wargentin was not yet finished – by developing systematic new ways to collect and interrogate data, he opened up further doors that still shape the modern world.

In partnership with Strömer in Uppsala, Wargentin undertook a massive government-sponsored project to survey the entire Baltic Sea coastline of Sweden. For this, he scaled up and applied the same triangulation methods that Celsius, de Maupertuis and their colleagues had employed in the Arctic twenty years earlier. The outcome was a vast improvement in nautical charts, upon which the nation's prosperity so much depended. Celsius' preoccupation with making science practically useful was in very safe hands.

Like Celsius, Wargentin also began making meteorological observations. From 1756, he logged daily temperature, air pressure and other weather events in Stockholm, thereby adding a valuable counterpoint to the existing Uppsala data series and others just beginning in other countries. As the information at his disposal grew, Wargentin made a startling discovery – arguably the world's first scientific evidence of climate change.

Instead of focusing on the extreme upper and lower temperatures (as recorded in Uppsala since 1722), Wargentin chose to record *average* temperatures. He took the mean figures over ten-day periods, calendar months and whole years, then compared them to his own Stockholm data and observations made in Paris. Unsurprisingly, this showed that winters in Paris were shorter and warmer than in Sweden, but also that winters in Scandinavia were becoming steadily longer and milder. Fascinated by this insight and evidence of the planet's dynamic nature, Wargentin wrote: 'We hope that the winter and spring cold will not continue to increase in the future, and that the seasons will return to their normal state.' As history has shown, he was wrong on both counts. When it comes to climate there is no normal, just never-ending flux.

Wargentin made one more breakthrough contribution from which it is possible to draw a straight line to today's climate concerns. In 1749, the Swedish Parliament commissioned him to collate and analyse all of the nation's records of births, deaths and marriages. Although this information was collected by the Church of Sweden, it was spread across thousands of cities, towns and parishes and had never been brought together or studied as a whole. With his aptitude for digesting and analysing huge amounts of information, Wargentin set about the task and combined it with another topic that had recently caught his attention: people's life expectancy.

For the second element, he drew on the actuarial calculations used by insurance companies. Once complete, the statistics created the world's first systematically assembled picture of a country's population and how it was changing. Census-taking was nothing new, of course – the Babylonians, Chinese dynasties and Egyptian kingdoms had practised it for thousands of years to administer armies, workforces, taxes and privileges. But Wargentin's study put demographics onto a different plane – one later taken up in England by the economist Thomas Malthus in his prophetically titled *Essay on the Principle of Population, or a View of Its Past and Present Effects on Human Happiness, with An Enquiry into Our Prospects Respecting the Future Removal or Mitigation of the Evils Which It Occasions.*[5]

Malthus held that growth in the world's population was potentially exponential, while the planet's resources and ability to feed ever larger numbers of people were more limited and linear. In graphical terms, Malthusian growth shows population doubling faster and faster up a steepling curve, while food supply falls progressively behind on a straight

The English economist Thomas Malthus (1766–1834), who warned of the dangers of human population growth. Portrait by John Linnell.

line. The Englishman warned that this must inevitably lead to catastrophe, where agricultural production collapses, leading to famine and war. Alarmed by these predictions, the British government moved rapidly to instigate its first national census in 1801, which in turn became the template for population studies throughout the world. Malthus' essays were a major influence on Charles Darwin in developing his theories of evolution.[6]

Modern real-time computer modelling of global population growth and forecasting owes a great debt to Wargentin's analysis of the patterns of life and death in eighteenth-century Sweden. His name is not so familiar as those who built upon his work, but fittingly, he is remembered in space, as a large crater on Earth's moon named after him. It lies just on the edge of the Moon's visible face, and it has a story of its own to tell.

In most impact craters, the floor is lower than the surrounding terrain, due to the punching force of whatever projectile creates them. But not so Wargentin – when you catch it in the right light, it is clear to see that the collision here penetrated far deeper, rupturing the surface so that molten lava rose and filled it to the wrinkle-edged brim. This impact, like the energetic scientist from whom the resulting crater takes its name, has left an indelible impression.

Of all the people on whom Celsius' influence rubbed off, it is perhaps his friend Linnaeus who he helped most to ascend to the top tier of scientific authority. After his 2,000-kilometre clockwise trip around the Gulf of Bothnia in 1732, Linnaeus' career barely looked back. His expedition journal records 100 previously unidentified species of plants and mentions him stopping one day and staring at the jawbone of a horse lying on the ground. In a flash of insight he remarked: 'If I only knew how many teeth and of what kind every animal had, how many teats and where they were placed, I should perhaps be able to work out a perfectly natural system for the arrangement of all quadrupeds.'[7]

Linnaeus' time in the Arctic formed the basis for his book *Flora Lapponica*[8] and the universal nomenclature that he went on to develop and publish in *Systema Naturae*,[9] *Philosophia Botanica*[10] and *Species Plantarum*.[11] Linnaeus' revolutionary approach to classifying all animals and plants bypassed the limitations and dead-ends of those who had tried to do the same thing before him, as far back as Aristotle, Theophrastus and Pliny the Elder. By grouping species into family trees with a clear lineage back to common ancestors, he was able to describe every organism with just two words rather than dozens. Hence, the honeybee, previously designated *Apis pubescens, thorace subgriseo, abdomine fusco, pedibus posticus glabis, untrinque margine ciliatus*, became *Apis mellifer* – a simple combination of genus and species made possible by the new terminology.

Other travel and expedition opportunities arose for Linnaeus. In 1735, he departed for the Dutch Republic to secure his doctorate and, by doing so, outsmart the scheming Browallius to win the hand of marriage to Sara Elisabeth Moraea. On his way to the Netherlands, he stopped in Hamburg, where the burgermeister proudly showed him what he claimed (and apparently believed) to be the taxidermised remains of a seven-headed hydra – the diabolical beast of Greek mythology. Linnaeus quickly saw that it was no more than the heads and claws of weasels covered in snakeskin, and he undiplomatically exposed the fake in public. The town official, embarrassed and frustrated perhaps at missing the opportunity to sell the specimen, made it clear that his visitor was no longer welcome in the city and ushered him on his way.

Linnaeus continued to trace a similar path to Celsius, visiting England in July 1736, where he met Sir Hans Sloane, and also stopping in Paris for a month on his return to Sweden. Back in Uppsala in 1738, he was appointed professor of medicine, which allowed him to continue restoring and developing Rudbeck's garden, where he had first met Celsius' uncle Olof. Linnaeus turned the space into a living textbook of flora and fauna, including free-roaming peacocks, monkeys living in tiny wooden houses perched on poles and a pet racoon. He even appointed the architect Carl Hårleman, who had worked on Celsius' observatory a few streets away, to design an elegant orangery at the end of the garden to cap his creation with classical splendour.

In 1750, Linnaeus was appointed rector of Uppsala University. He became wealthy enough to acquire two country estates outside Uppsala, where he relocated most of his library and collections in case of another fire in the city. He died in 1778 and lies at the entrance to Uppsala Cathedral, his name deeply carved into a dark stone slab. He is still present throughout the city, not just at the garden that provides citizens and tourists with a relaxing green sanctuary, but in the names of dozens of streets, shops, cafés and restaurants. Uppsala is rightly proud of this famous son, the father of systematic biology. And fittingly, on the front façade of the main university building, Linnaeus and Celsius are memorialised side by side, together in death as they were in life.

Linnaeus' contemporaries and disciples heaped praise upon him. The Geneva-born philosopher, writer and composer Jean-Jacques Rousseau sent him the message: 'Tell him I know no greater man on Earth', while Johann Wolfgang von Goethe wrote: 'With the exception of Shakespeare and Spinoza, I know no one among the no longer living who has influenced me more strongly.' In the following century, the Swedish playwright, novelist and essayist August Strindberg reflected: 'Linnaeus was in reality a poet who happened to become a naturalist.' As the ultimate and lasting symbol of his importance, Linnaeus' body is still formally classified as the internationally recognised type specimen for human beings. He remains the singular embodiment of the two-word species name he invented: *Homo sapiens*.

Celsius had no children, but his family name also continued to resonate in and around Uppsala. His cousin Olof Celsius the Younger (1716–94) became the Bishop of Lund and later deputy librarian and professor of history at Uppsala. Olof was knighted in 1756, entitling him and his nine children to use the ennobled name von Celse, and leading to a successful career in Hats Party politics and the civil service. From rustic and meagre beginnings, the Celsius family had risen to the heights implicit in its name.

Linnaeus' 1735 categorisation of the animal kingdom.

REGNUM ANIMALE.

IV PISCES.

Corpus apodum, pinnis veris instructum, nudum, vel squamosum.

Order	Genus	Character	Species
PLAGIURI. *Cauda horizontalis.*	Thrichechus.	*Dentes* in utraque maxilla. *Dorsum* impenne.	Manatus s. *Vacca mar.*
	Catodon.	*Dentes* in inferiore maxilla. *Dorsum* impenne.	Cet. Fistula in rostro *Art.* Cete *Clus.*
	Monodon.	*Dens* in superiore max. 2. *Dorsum* impenne.	Monoceros. *Unicornu.*
	Balæna.	*Dentes* in sup. max. cornei. *Dorsum* sæpius impenne.	B. Groenland. B. Finfisch. B. Maxill. inf. latiore. *Art.*
	Delphinus.	*Dentes* in utraque maxilla. *Dorsum* pinnatum.	Orcha. Delphinus. Phocæna.
CHONDROPTERYGII. *Pinnæ cartilagineæ.*	Raja.	*Foramina* branch. utrinq. 5. *Corpus* depressum.	Raja clav. asp. læv. &c. Squatino-Raja. Altavela. Pastinaca mar. Aquila. Torpedo. Bos *Vet.*
	Squalus.	*Foram.* branch. utrinq. 5. *Corpus* oblongum.	Lamia. Galeus. Catulus. Vulpes mar. Zygæna. Squatina. Centrine. Pristis.
	Acipenser.	*Foram.* branch. utrinq. 1. *Os* edentul. tubulatum.	Sturio. Huso. *Ichthyocolla.*
	Petromyzon.	*Foram.* branch. utrinq. 7. *Corpus* bipenne.	Enneophthalmus. Lampetra. Mustela.
BRANCHIOSTEGI. *Pinnæ offic. carentes. Branch. off. & membran.*	Lophius.	*Caput* magnitudine corporis. *Appendices* horizontaliter latera piscis ambiunt.	Rana piscatrix. Guacucuja.
	Cyclopterus.	*Pinnæ* ventrales in unicam circularem concretæ.	Lumpus. *Lepus mar.*
	Ostracion.	*Pinnæ* ventrales nullæ. *Cutis* dura, sæpe aculeata.	Orbis div. sp. Pisc. triangul. Atinga. Hystrix. Ostracion. Lagocephalus.
	Balistes.	*Dentes* contigui maximi. *Aculei* aliquot robusti in dorso.	Guaperua. Histrix. Capriscus. *Caper.*
ACANTHOPTERYGII. *Pinnæ osseæ, quarum quædam aculeatæ.*	Gasterosteus.	*Membr. branch.* officula 3. *Venter* laminis osseis instr.	Aculeatus. Spinachia. Pungitius.
	Zeus.	*Corpus* compressum. *Squamæ* subasperæ.	Aper. Faber. *Gallus mar.*
	Cottus.	*Membrana branch.* offic. 6. *Caput* aculeatum, corpore latius.	Cataphractus. Scorpio mar. Cottus. *Gobio fl. capit.*
	Trigla.	*Appendices* ad pinn. pect. articulatæ 2 vel 3.	Lyra. Gurnardus. Cuculus. Lucerna. Hirundo. Milvus. Mullus barb. & imberb.
	Trachinus.	*Opercula branch.* aculeata. *Oculi* vicini in vertice.	Draco. Araneus mar. Uranoscopus.
	Perca.	*Memb. branch.* officul. 7. *Pinnæ* dorsales 1 vel 2.	Perca. Lucioperca. Cernua. Schraitser.
	Sparus.	*Opercula branch.* squamosa. *Labia* dentes tegunt. *Dentes* molares obtinet.	Salpa. Melanurus. Sparus. Sargus. Chromis. Mormyrus. Mæna. Smaris. Boops. Dentex. Erythrinus. Pagrus. Aurata. Cantharus.
	Labrus.	*Labia* crassa dentes teg. *Color* speciosus.	Julis. Sachetrus. Turdus diversar. specier.
	Mugil.	*Memb. branch.* offic. 6. *Caput* totum squamosum.	Mugil. *Cephalus.*
	Scomber.	*Memb. branch.* offic. 7. *Pinnæ dorsi* 2 vel plures.	Glaucus. Amia. Scomber. Thynnus. Trachurus. *Saurus.*
	Xiphias.	*Rostrum* apice ensiformi. *Pinnæ ventrales* nullæ.	Gladius.
	Gobius.	*Pinnæ vent.* in 1 simpl. concr. *Squamæ* asperæ.	Gob. niger. Jozo. Paganellus. *Aphua.*
MALACOPTERYGII. *Pinnæ osseæ, quæ omnes molles.*	Gymnotus.	*Membr. branch.* officul. 5. *Pinna* dorsalis nulla.	Carapo.
	Muræna.	*Membr. branch.* offic. 10. *Tubuli* in apice rostri. 2.	Anguilla. Conger. Fluta. Serpens mar.
	Blennus.	*Pinnæ ventr.* constant off. 2. *Caput* admodum declive.	Alauda non crist. & galer. Blennus. Gattorugine.
	Gadus.	*Membr. branch.* offic. 7. *Pinnæ* dorsi 2 vel 3.	Asellus diversar. specier. Merluccius. Anthias adus. Mustela. Egrefinus.
	Pleuronectes.	*Membr. branch.* off. 6. *Oculi* ambo in eodem later.	Rhombus divers. specier. Passer. Limanda. Hippoglossus. Bugloss. Solea.
	Ammodytes.	*Membr. branch.* offic. 7. *Pinnæ* ventr. nullæ.	Ammodytes. *Tobianus.*
	Coryphæna.	*Membr. branch.* offic. 5. *Pinna* dorsi a capite ad caudam.	Hippurus. Pompilus. Novacula. *Pesten.*
	Echeneis.	*Stria* transversæ, asperæ, in superna capitis parte.	Remora.
	Esox.	*Membr. branch.* offic. 14.	Lucius. Belone. Acus maxima squamosa.
	Salmo.	*Membr. branch.* offic. 10-12. *Corpus* maculosum.	Salmo. Trutta. Umbla. Carpio lacustr.
	Osmerus.	*Membr. branch.* offic. 7--8. *Dentes* in max. lingu. palat.	Eperlanus. *Spirinchus.* Saurus.
	Coregonus.	*Membr. branch.* offic. 8--10. *Appendix* pinniformis.	Albula. *Lavaretus.* Thymallus. Oxyrhynchus.
	Clupea.	*Membr. branch.* offic. 8. *Venter* acutus serratus.	Harengus. Spratti. Encrasicholus. Alosa.
	Cyprinus.	*Membr. branch.* offic. 3. *Dentes* ad orificium ventriculi tantum.	Erythrophthal. Mugil. *fluv.* Brama. Ballerus. Capito. Nasus. A. M. Carassius. Cypr. nobilis. Tinca. Barbus. Rutilus. Alburnus. Leuciscus. Phoxinus. Gobius fl.
	Cobitus.	*Caput* compressum. *Pinna* dorsi & ventrales eidem a rostro distantia.	Cobitis. Barbatula. Misgurn.
	Syngnathus.	*Opercula branch.* ex laminis 2. *Maxilla* a lateribus clausæ.	Acus lumbr. Acus *Aristot.* Hippocampus.

V. INSECTA.

Corpus crusta ossea cutis loco tectum. *Caput* antennis instructum.

Order	Genus	Character	Species
COLEOPTERA. *Alæ elytris duabus tectæ.*	Blatta.	§. FACIE EXTERNA FACILE DISTING. *Elytra* concreta. *Alæ* nullæ. *Antennæ* truncatæ.	Scarab. tardipes. Blatta fœtida.
	Dytiscus.	*Pedes* postici remorum forma & usu. *Ant.* setaceæ. *Sterni* apex bifurcus.	Hydrocantharus. Scarab. aquaticus.
	Meloë.	*Elytra* mollia, flexilia, corpore breviora. *Ant.* moniliformes. *Ex* articulis oleum fundens.	Scarab. majalis. Scarab. unctuosus.
	Forficula.	*Elytra* brevissima, rigida. *Cauda* bifurca.	Staphylinus. Auricularia.
	Notopeda.	Pectus in dorso *exsilit.* *Ant.* capillaceæ.	Scarab. elasticus.
	Mordella.	*Cauda* aculeo rigido simplici armata. *Ant.* setaceæ, breves.	Negatur ab *Aristotele.*
	Curculio.	*Rostrum* productum, teres, simplex. *Ant.* clavatæ in medio Rostri positæ.	Curculio.
	Buceros.	*Cornu* 1. simplex, rigidum, fixum. *Ant.* capitatæ, foliaceæ.	Rhinoceros. Scarab. monoceros.
	Lucanus.	*Cornua* 2. ramosa, rigida, mobilia. *Ant.* capitatæ, foliaceæ.	Cervus volans.
	Scarabæus.	§. ANTENNÆ TRUNCATÆ. *Ant.* clavatæ foliaceæ. *Cornua* nulla.	Scarab. pilularis. Melolontha. Dermestes.
	Dermestes.	*Ant.* clavatæ horizontaliter perfoliatæ. *Clypeus* planiusculus, emarginatus.	Cantharus fasciatus.
	Cassida.	*Ant.* clavato-subulatæ. *Clypeus* planus, antice rotundatus.	Scarab. clypeatus.
	Chrysomela.	*Ant.* simplices, clypeo longiores. *Corpus* subrotundum.	Cantharellus.
	Coccionella.	*Ant.* simplices, brevissimæ. *Corpus* hemisphæricum.	Cochinella vulg.
	Gyrinus.	*Ant.* simplices. *Corpus* breve. *Pedibus* posticis saliens.	Pulex aquaticus. Pulex plantarum.
	Necydalis.	*Ant.* clavato-productæ. *Clypeus* angustus, rotundatus.	Scarabæo-formica.
	Attalabus.	*Ant.* simplices, constructæ articulis orbicularibus, præter ultim. globosum.	Scarab. pratensis.
	Cantharis.	§. ANTENNÆ SETACEÆ. *Clypeus* planus, margine undique promin. *Elytra* flexilia.	Cantharis *offic.*
	Carabus.	*Clypeus* fere planus, marg. prominente. *Elytra* fragilia.	Cantharus fœtidus. Cantharellus auratus.
	Cicindela.	*Clypeus* cylindraceus vel teres. *Forceps* oris prominens.	Cantharus Marianus.
	Leptura.	*Clypeus* subrotundus. *Pedes* longi. *Corpus* tenue acuminatum.	Scarab. tenuis.
	Cerambyx.	*Clypeus* ad latera mucrone prominet. *Ant.* corpus longitudine æquant, vel superant.	Capricornus.
	Buprestis.	*Clypeus* superne a punctis elevatis notatus.	Scarab. sylvaticus.
ANGIOPTERA. *Alæ omnibus datæ, elytris destitutæ.*	Papilio.	*Rostrum* spirale. *Alæ* 4.	Papilio alis erectis. Psyche --- planis. Phalæna --- compressis.
	Libellula.	*Cauda* foliosa. *Alæ* 4. expansæ.	Perla. Virguncula.
	Ephemera.	*Cauda* setosa. *Alæ* 4. erectæ.	Musca Ephemera.
	Hemerobius.	*Cauda* setosa. *Alæ* 4. compressæ.	Phryganea.
	Panorpa.	*Cauda* cheliformis. *Alæ* 4. *Rostr.* corn.	Musca scorpiurus.
	Raphidia.	*Cauda* spinoso-setacea. *Alæ* 4. *Cap.* corn.	*Non descr.*
	Apis.	*Cauda* aculeo simplici. *Alæ* 4.	Crabro. Vespa. Bombylius. Apis.
	Ichneumon.	*Cauda* aculeo partito. *Alæ* 4.	Ichneumon. Musca tripilis.
	Musca.	*Stylus* sub alis capitatus. *Alæ* 2.	Muscæ div. spec. Oestrum *Vet.* Oestrum *Lapponum.* Tabanus. Culex. Teredo nav. Tipula. Formica-leo.
HEMIPTERA. *Alæ elytris destitutæ, Quibusdam tantum individuis concessæ.*	Gryllus.	*Pedes* 6. *Alæ* 4. *superiores* crustaceæ.	Gryllus domesticus. Gryllo-talpa. Locusta. Mantis.
	Lampyris.	*Pedes* 6. *Clypeus* planus. *Alæ* 4.	Cicindela.
	Formica.	*Pedes* 6. *Alæ* 4. *Cauda* aculeum condit.	Formica.
	Cimex.	*Pedes* 6. *Alæ* 4. cruciferæ. *Rostrum* styliforme, rectum.	Cimex lectularius. Orsodachne. Tipula aquatica. Bruchus.
	Notonecta.	*Pedes* 6. quorum *postici* remorum figura & usu. *Alæ* 4. cruciferæ.	Notonecta aquatica.
	Nepa.	*Pedes* 4. *Frons* chelifera. *Al.* 4. crucif.	Scorpio aquat.
	Scorpio.	*Pedes* 8. *Frons* chelifera, aculeata. *Alæ* 4. laxæ.	Scorpio terrestris.
APTERA. *Alæ nullæ.*	Pediculus.	*Pedes* 6. *Antennæ* capite breviores.	Pediculus humanus. Ped. avium. Ped. piscium. Ped. pulsatorius.
	Pulex.	*Pedes* 6. saltatrices.	Pulex vulgaris.
	Monoculus.	*Pes* 1. *f. Antennæ* bifidæ.	Pulex arboresc. *Swam.* *Monoculus Bradl.* *Apus Frisch.*
	Acarus.	*Pedes* 8. articulis 8 constantes. *Oculi* 2. *Ant.* minimæ.	Ricinus. Scorpio-araneus. Pedic. inguinalis. Pedic. Scarabæi. Pedic. Scabiei. Araneus coccineus.
	Araneus.	*Pedes* 8. *Oculi* communiter 8.	Araneus. Tarantula. Phalangium.
	Cancer.	*Pedes* 12. *priores* cheliformes.	Cancer. Astacus. Pagurus. Squilla. Majas. Eremita. Gammarus.
	Oniscus.	*Pedes* 14.	Asellus Offic. Asellus aquat.
	Scolopendria.	*Pedes* 20. & ultra.	Scolop. terrestris. Scolop. marina. Julus.

VI. VERMES.

Corporis *Musculi* ab una parte basi cuidam solidæ affixi.

Order	Genus	Character	Species
REPTILIA. *Nuda, artubus destituta.*	Gordius.	*Corpus* filiforme, teres, simplex.	Seta aquatica. Vena Medina.
	Tænia.	*Corpus* fasciatum, planum, articulatum.	Lumbricus longus.
	Lumbricus.	*Corpus* teres, annulo prominenti cinctum.	Intestinum terræ. Lumbricus latus. Ascaris.
	Hirudo.	*Corpus* inferne planum, superne convex. tentaculis destitutum.	Sanguisuga.
	Limax.	*Corpus* inferne planum, superne conv. tentaculis instructum.	Limax.
TESTACEA. *Habitaculo Lapideo instructa.*	Cochlea.	*Testa* univalvis, spiralis, unilocularis.	Helix. Labyrinthus. Voluta. Cochlea varia. Buccinum. Lyra. Turbo. Cassida. Strombus. Fistula. Terebellum. Murex. Purpura. Aporrhais. Nerita. Trochus.
	Nautilus.	*Testa* univalvis, spiralis, multilocularis.	Nautilus. Orthoceros. Lituus.
	Cypræa.	*Testa* univalvis, convoluta, rima longitudinali.	Concha Veneris. Porcellana.
	Haliotis.	*Testa* univalvis, patula, leviter concava, perforata, ad angulum spiralis.	Auris marina.
	Patella.	*Testa* univalvis, concava, simplex.	Patella.
	Dentalium.	*Testa* univalvis, teres, simplex.	Dentalium. Entalium. Tubus vermicul.
	Concha.	*Testa* bivalvis.	Mytulus. Vulva marina. Pholas. Bucardium. Perna. Chama. Solenes. Tellina. Pinna. Ostrea. Pecten. Mitella. *Plinii.*
	Lepas.	*Testa* multivalvis. *Valvulæ* duabus plures.	Concha anatifera. Verruca testudin. Balanus marinus.
ZOOPHYTA. *Artubus donata.*	Tethys.	*Corpus* forma variabile, molle, nudum.	Tethys. Holothurium. Penna marina.
	Echinus.	*Corpus* subrotundum, testa tectum, aculeis armatum.	Echinus marinus.
	Asterias.	*Corpus* radiatum, corio tectum, scabrum.	Stella marina. Stella oligactis. St. pentactinoides. St. polyactinoid.
	Medusa.	*Corpus* orbiculatum, gelatinosum, subtus filamentosum.	Urtica marina. Urt. vermiformis. Urt. crinita. Urt. astrophyta.
	Sepia.	*Corpus* oblongum, interne osseum, anterius octo artubus donatum.	Sepia. Loligo.
	Microcosmus.	*Corpus* variis heterogeneis tectum.	Microcosm. marin.

PART VI

TIME

⧗

22

A SAFE AND JUST EARTH FOR ALL

Celsius' Legacy, our Indifferent Planet and Hope

There is no soft landing for humanity.
> Professor Dr Johan Rockström, Potsdam Institute for Climate Impact Research, Uppsala Celsius Lecture, February 2023

'Here emerges our urban community garden' read the modest sign attached to a tree in a shady green space outside Geocentrum, Uppsala University's Department for Earth Sciences. I was on my way to meet Professor Tom Stevens, a British expert in – as he's proud to say – dust. More specifically, he studies wind-blown loess deposits in China and eastern Europe to understand the interaction of Earth's climate, surface processes and tectonics. Tom's research spans the Cenozoic-Quaternary – the past 2.58 million years. Such a period is hard to contemplate and visualise, and harder still when you recall this does not even amount to one-thousandth of all geological time.

The sign explained that the fenced area is part of a student-led project to 'engage a diverse crowd, generating new relationships and cross learning while creating a safe outdoor space for interaction'. The garden, it said, was 'inspired by permaculture principles and circular system designs' with a goal 'to integrate the garden with existing ecosystems while reusing materials and creating solutions that minimize waste'. 'Everyone is welcome!' it concluded in bold text.

I stopped and looked at the sign and the untamed garden beyond. Here, I thought, was something wonderful: a glimpse of what the world could and should be, a microcosm of the choices facing humanity in the twenty-first century. Inside this little patch of campus is a vision for somewhere people can live and prosper differently, in much greater harmony with the planet than we have managed so far. The permaculture invoked here is about Earth care, people care and fair shares – a radically different approach to the belief systems that have dominated the last two millennia. Permaculture blends ancient indigenous practices with cutting-edge ecological science and design.[1]

The optimism of the garden and sign stayed with me as I met Tom and his Swedish colleague, Christoffer Hallgren. They took me a short way past the main Geocentrum building to a circular open space surrounded by galvanised steel railings. The grassy circle was no more than 10 metres across, and inside stood an aluminium mast festooned with monitoring equipment. As Christoffer explained the functions of the various devices, an anemometer whizzed round at the top of the mast, recording the wind speed. Inside the compound there were lasers to measure the density and form of precipitation, radiation sensors and well-worn rain gauges.

Beside the mast stood a narrow metal pipe about 1 metre high, marked in centimetre graduations. It seemed to come from a different age to the rest of the equipment. 'And what's this?' I asked, tapping the rusty top of the pipe. 'Ah – that's how we measure the depth of the snow', said Christoffer.

It was a comic but precious moment – a reminder that even for sophisticated academic brains, the simple solutions are sometimes still the best. It was a special exchange too, because we were standing in the space now devoted to continuing the Uppsala Weather Series – the daily records begun in 1722 by Burman and Celsius and maintained ever since. I looked at the two modern-day scientists, relaxed in their shorts, tee-shirts and sunglasses, and thought how different they were to their frock-coated and tricorn-hatted predecessors. Or perhaps not so different, since they are engaged in exactly the same enterprise as Celsius and his peers: to understand, analyse and predict patterns in the Earth's climate and what these mean for all of its inhabitants. Christoffer was pleased to tell me that, like Celsius, he is 'Uppsala born and bred'.

⌛

I had set out to discover more about Celsius – the unknown man with the well-known name – and what lessons he and his work might hold for current generations and those to come. Over lunch with Tom in the nearby botanical gardens, I started to take stock of what I thought these lessons were.

Celsius' and Burman's first page of entries for the Uppsala Weather Series in 1722. Its daily measurements still continue today.

The first point, as my late father often used to say about many things, is that 'it's all about TIME'. We cannot hope to appreciate and respond appropriately to what is happening where we live without a proper conception of the enormous timescales involved in forming and shaping the Earth. After four and a half billion years, our Sun is roughly half-way through its expected lifespan. When the Sun eventually dies, it will first grow into a fiery red giant, thousands of times its current size. This will engulf us and the other closest planets, but then shrink to a super-dense, Earth-sized white dwarf – part of the continuous recycling, transformation and expansion of cosmic matter that makes up the universe. And just as the universe exploded into existence around 13.8 billion years ago, it seems likely that it too will eventually reverse, collapse, squeeze or freeze – and cease to be.

One theoretical prediction of this state envisages all matter being crushed into an infinitely dense singularity, billions or trillions of years hence. And in a nod to the work of Celsius and Kelvin, it seems that the behaviour of atoms as they approach absolute zero will be a determining feature of this stage.[2] Scientific opinion is divided on whether this will then trigger another Big Bang and new universe, or whether it will simply be the final curtain for everything.

Ingenious individuals over many centuries have discovered and formulated this knowledge, but history suggests that, as a species, we are ill-equipped to readily understand and make use of it. Celsius, though, was an exception. In everything he did, he demonstrated the ability to think long term – constantly asking questions about the past and present with a firm eye to the future. And within this lie two other important lessons.

Given the immensity of time, it is easy to view the world around us as being in a steady state – things as they are now being much as they always were, and always will be. But this is completely wrong. Everything in the universe is dynamic, constantly changing and interacting, even if we cannot perceive it. This in turn means that there is no limit on what we are yet to experience or discover – forces, effects and events that are hidden, just emerging, or which lie ahead. Celsius seemed to intuitively grasp and work with this reality in ways that many of his predecessors and successors did not. He understood that theories, hypotheses, experiments and results are often, at best, explanations of phenomena. They are rarely the full story, so they should always remain open to review and new or contrary evidence.

Celsius' intellectual ease with time reveals yet another fundamental principle – something that humans must contemplate as we decide how to live in the decades and centuries to come. Much of Celsius' work focused on how tiny, successive actions can lead to huge, world-changing outcomes. From the countless subatomic charged particles that create the Northern Lights, to the sea and land changes in the Baltic, or the extraction of natural resources at the mines in Sala and Falun, Celsius looked at natural and human-made systems and saw connections. What each of us does today, tomorrow, the day after and the day after that dictates how the world will be in aeons' time. And this awareness creates responsibilities and accountability for our actions.

Today we are living with the consequences of industrialisation in past centuries and decades. Much of the global warming forecast up to 2050 is already irreversibly locked into future climate patterns.[3] Similarly, it will take hundreds, thousands or millions of years to recoup the acidification, deforestation, soil loss and depletion of resources that has already occurred. What is done is done, but we can decide on how we behave now and into the future.

Sitting in the Uppsala botanical gardens, I also reflected on how the drama of the European Enlightenment produced hundreds of outstanding thinkers, scientists and artists – men and women who questioned the status quo and sought to make things better. Cast members like Celsius and his friend Linnaeus demonstrated what brilliant individuals are capable of. But their lives also illustrate the limitations of purely personal action. Positive, lasting change demands not just exceptional people; it also requires cultures and systems within which they can operate. In retrospect at least, the Enlightenment was an example of just such a framework. It was a time that celebrated knowledge, encouraged learning and rewarded pro-social endeavour. But it was a delicate season soon ended by political and mercantile interests.

Finding solutions to the interconnected polycrisis of wars, poverty, inequality, disease, ecological collapse and climate change now affecting global society will need more than supremely gifted people, no matter how inspirational or courageous they might be. There seems little evidence that the established network of international institutions, nation states, governance, faiths, markets or standard human discourse are up to the job of facing such existential challenges. In fact, as Tom pointed out as we ate our lunch in the sunny garden café, it would be hard to design a world system more guaranteed to bring about conflict and catastrophe than what we have now. If *Homo sapiens* wish and choose to survive for the next few centuries and have a place to enjoy after that, we will need new Enlightenments and new Renaissances to bring out and apply the best of ourselves.

By following in Celsius' footsteps, from his birthplace and tomb to the crowning achievement of his career inside the Arctic Circle, I learned that his lasting relevance is about much more than science. It is his modus operandi and habits that count just as much. He was an instinctive collaborator and audacious internationalist – generous in sharing his ideas and resources, eager to learn from others and never afraid to go into unfamiliar places and circumstances. When vision, inspiration and courageous leadership were called for, Celsius could provide it. But he could also take direction, cooperate and follow others, confident in his own abilities but open to informed opinions.

Celsius made his name and left his mark not by domineering or aggressive behaviour and overcoming his competitors; he went about his studies, travels and science in a gentle and persuasive fashion. He was also driven, canny and, on occasion, capable of sharp tactics – like the secret commissioning of expensive instruments for which his university was then compelled to pay. But for the most part, he lived and worked with care for those around him and respect for the natural world that so excited his interest.

Celsius showed that in science, and every aspect of life, manners matter. The relationships we form and conventions we observe are the glue that holds civilisation together. If we continue to neglect or abandon these safeguards and treat our place in the universe as a zero-sum game we need to win, we face a disorderly and imminent self-annihilation. What is needed is something closer to Émile Durkheim's concept of 'collective effervescence' – the sacred state in which a community unites around a shared thought, principle or action.[4] The French philosopher recognised the need for humans to have a totem – something they can share and communicate to unlock mutual excitement, energy and possibility. In a mood that Celsius would surely have welcomed, in 1912 he wrote:

> Collective consciousness is the highest form of the psychic life, since it is the consciousness of the consciousnesses. Being placed outside of and above individual and local contingencies, it sees things only in their permanent and essential aspects, which it crystallizes into communicable ideas. At the same time that it sees from above, it sees farther; at every moment of time, it embraces all known reality; that is why it alone can furnish the mind with the moulds which are applicable to the totality of things and which make it possible to think of them. It does not create

these moulds artificially; it finds them within itself; it does nothing but become conscious of them.[5]

There is no question that humans, with all our ingenuity and resilience, are capable of this order of thinking. What is in doubt is whether we will choose this path, especially at a time when we are nearing, or have possibly already reached, multiple environmental tipping points. The climate crisis facing us now is outside anyone's experience; and it seems we lack the language, narratives and reference points to comprehend it properly or agree on the direction of change. It is not just a question of things becoming slightly hotter, colder, drier or wetter, but a complete disruption of the natural cycles to which we are accustomed. There is strong evidence and plenty of history to show that, when critical stages like this are reached, change can happen very quickly – far faster than societies are able to cope with.

Back in Tom's office after lunch, he showed me a graph of abrupt climatic changes during the Earth's last glacial period (roughly 115,000– 11,700 years ago), based on analysis of ice core samples collected in Greenland. The columns and lines traced the correlation between the dust particles in the samples and the atmospheric gases to which they had been exposed and average temperatures. Tom pointed to some of the spikes in the timeline. They showed that when a dust peak occurs (say from volcanic activity, an asteroid strike, human-made pollution or nuclear explosion) temperatures can alter very dramatically and suddenly – sometimes within a few years or decades.

On a sparsely populated planet, with abundant natural resources, humans have been able to survive this sort of event in the past. But on an Earth denuded by ecocide and with a human population set to pass 10 billion by 2100, widespread survival will be much more difficult. Tom put his finger on one of the graph's extremes and said: 'Policy has to be based on the knowledge that this can and will happen.'

⧖

Walking around Uppsala today, it is striking how small a radius Celsius occupied for most of his life. His birthplace, university quarters and observatory all lie within a few hundred metres of each other, and his grave is just a short distance outside the city. Apart from his European travels and expeditions in 1732–37, Celsius spent his entire life within a tiny portion of the world, from which he was able to form the biggest ideas and conjectures. He was an original type specimen for the 'think global, act local' mindset advocated by many who are anxious to conserve the Earth's diversity and potential.

Celsius' visits to Sweden's silver and copper mines would have confronted him with the fact that the history of human civilisation has been based on extracting and exploiting minerals. There are those we use for manufacturing, consumption or their perceived value (such as coal, oil, iron, copper, silicates, silver and gold), and those we apply indirectly to support our diets and agriculture (like phosphates, sulphates, potassium and calcium). Up to now, our species has multiplied and spread by treating the natural environment as a giant, interest-free bank – a reservoir of resources that provides all the space and all the raw materials and bears all the cost of our activity. Central to the questions confronting humanity now is whether it has to be that way, or whether we can pursue different paths to survival.

I like to imagine that, if they were alive today, big thinkers like Celsius and Linnaeus would be busily engaged in this issue, their penetrating intellects taking the debate to fresh and interesting places. Could the whole basis of how we live be inverted, so that, rather than the environment perpetually bankrolling human activity and taking the hit for our excesses, it becomes the prime beneficiary of what we do and the target for the best things of which we are capable?

The modest list of Celsius' belongings, which his mother compiled after his death, demonstrates that he was not motivated by or especially concerned about material wealth. Might this be a cue for a new and different human era, in which economic growth becomes supplanted by planetary stewardship and environmental regeneration as our guiding principles? Could economies be built on ideas, intelligence and creativity rather than removal, production and consumption? And what would it take to reach new settlements about this, both between ourselves and with the planet? Logic suggests that to succeed in this regard would demand thousands of solutions, sustained over centuries. But, in theory at least, it feels possible. The prize would be to recast human identity, extending it beyond self-gratification, acquisition and consumption into radically different cultural forms.

Moves to create a circular global economy, based on repair, reuse and recycling of resources, point firmly in this direction.[6] The model aims to replace current linear 'take-make-throw-away' habits with a sustainable approach that eliminates waste, repurposes products and allows the planet to begin a recovery. It recognises that the economy is a wholly owned subsidiary of the environment, not – as we have tended to view it and behave in the past – the other way around. Progressive experiments are under way to test this notion – for example, the city of Amsterdam has pledged to become a circular economy by 2050.[7]

Other initiatives seek to address spiralling over-consumption. In Switzerland, the 2000-Watt Society[8] movement has shown how joined-up governance and binding personal contracts can dramatically reduce the amount of energy used and carbon emitted. Living and working within an instantaneous use limit of 2,000 watts (2kW) per year per person assumes that three-quarters of this energy comes from renewable sources. This would bring lifestyles back within reasonable planetary boundaries and cut annual individual carbon emissions to around 1 tonne of CO_2. For comparison, in the United States the average person currently consumes around 12,000 watts (12kW) and emits 20 tonnes of CO_2 per year. Transitions of this sort are difficult certainly, but they also create new possibilities.

There are other exciting new technologies with the potential to completely reinvent whole sectors of the economy, particularly agriculture and food, which account for almost 40 per cent of global land use and a third of carbon emissions.[9] For example, widespread use of precision fermentation – growing microbes to produce edible protein – could cut the amount of land devoted to pasture animals by up to 80 per cent, freeing this up for other, less-damaging forms of food production.[10]

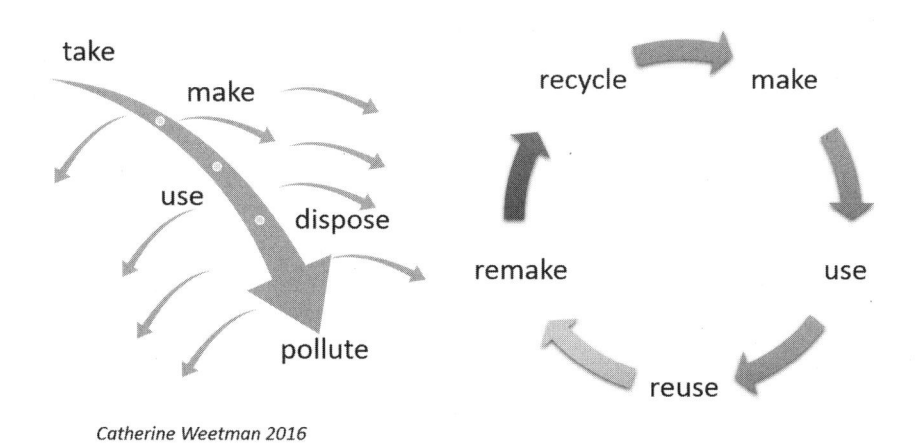

Catherine Weetman 2016

Can humans switch from linear production, consumption and waste to a more sustainable, circular global economy?

What should our attitude be towards alternatives and possibilities like these? Perhaps, like the eighteenth-century silver miners at Sala, we should see even a 50/50 chance of success as an opportunity worth pursuing. Great challenges like these require the openness, willingness to take calculated risks and ability to influence that Celsius displayed throughout his life. They also hold out the promise of a more communitarian future, in which, as individuals, we can become part of something much bigger than ourselves.

⌛

Just inside the entrance to Uppsala Cathedral, the walls of a side chapel are draped with a colourful tapestry depicting key events in the city's history. In one corner, it shows the building surrounded by smoke and flames in the terrible fire of 1702. Alongside, the standing figure of Olof Rudbeck looks on, eyes wide and beseeching hands held out in horror at the sight before him, his long grey hair streaming backwards from the force of the blaze.

As we progress through the twenty-first century, and hopefully beyond, will humans be like Rudbeck – brilliant but deluded, appalled at the destruction happening around us, wanting it to stop but powerless to fight the flames? Or will we be more like Celsius – able to stand back from the present and learn from the past? Are we condemned to just sit and await our fate or can we summon the will to leave outmoded orthodoxies behind and create a better future?

We are lucky organisms on a lucky planet. We have a choice.

EPILOGUE

ANDERS CELSIUS' MESSAGE TO THE TWENTY-FIRST CENTURY

The only way to escape the abyss is to look at it, gauge it, sound it out and descend into it.

Attrib. Cesare Pavese (1908–50)

During his Grand Tour visit to Bologna in 1733, a curious inscription upon an ancient tombstone about two lovers, Laelia and Agatho, caught Celsius' attention. He became fascinated by it and was moved to compose an imaginary ending to their tragic tale.[1]

Almost three centuries later, I was fortunate to research and write most of this book through a visiting scholarship at Wolfson College, University of Oxford. One day, while working in the library there, I came across and read a piece by the College's founding president, British philosopher Sir Isaiah Berlin. Composed against the backdrop of his own life, his 1994 *Message to the Twenty-First Century* is a warning against the dangers of fanaticism and of any credo that purports to hold a single, incontestable truth. In it he wrote:

> If you are truly convinced that there is some solution to all human problems, that one can conceive an ideal society which men can reach if only they do what is necessary to attain it, then you and your followers must believe that no price can be too high to pay in order to open the gates of such a paradise.[2]

Here I indulge a fantasy that fuses these two pieces of writing. This epilogue imagines that if, before his death, Celsius had also composed a message to our own times. What might he have said and wished for humankind now?

Like Celsius, Berlin recognised the complexity of reality. He saw the need to retain humility, pursue compromise and seek out new ideas that question orthodoxy and frustrate authoritarianism. As part of de Maupertuis' expedition to the Arctic in 1736–37, Celsius observed the stars to determine the shape of the Earth. His broader legacy is to help us understand and respond to the shape of our future.

◎

Dear friends

In the whole spectrum of natural creation, our species has unique capabilities. We can imagine, question, hypothesise, deduce and adapt like no other creature. While the deepest patterns and secrets of the universe may remain forever hidden to us, we can wonder, observe, explore, interpret and tell stories of all that we think and experience. This, I believe, is our purpose. Humans exist to <u>discover</u> and to <u>share</u>. And progress, betterment and change are the scales against which we measure success.

Our powers of imagination are so strong that we, in great part, organise our lives and relationships around things that do not — in a purely physical sense — exist. We are governed by faiths, customs, cultures, ideologies, ethics, money and states, but these are all abstract concepts conjured and crafted by our enquiring minds. We impose such constructs upon the world around us so as to bring order. But we should not confuse these guiding norms with what is <u>real</u>, and we must be always vigilant about the dangers of following conventions and ritual without question.

For humanity to thrive and be at peace, we should be mindful that much of what we do — our industry, our pleasures, our art and our trade — depend upon what <u>actually</u> exists in the natural treasury of the environment. And if we exhaust or deplete these resources to excess, we may perish. We should therefore be directed to preserving everything that is real and vital — the flora, fauna, minerals, seas and atmosphere of our planet. Nature is both the wellspring of life and the stream we should follow.

Throughout my work and travels I have benefited from cooperation between peoples of different regions, countries, races and creeds. Many of my own advances in science have been made possible by the generosity of hosts and colleagues from backgrounds and traditions quite different to my own. Again, the ever-shifting boundaries of nations that we draw upon globes and maps are but invented lines to help us make sense of the places we inhabit. These divisions are neither permanent, natural nor supernaturally ordained. They are just momentary descriptions and features of a single, much greater whole.

As an astronomer, I take a similar view about what lies beyond our planet and solar system. In the future, humanity's innate thirst to explore may lead us there, or entice us to send signals and envoys to distant realms. But we must be sure that our influence travels responsibly and treads lightly. The cosmos is indifferent to our existence, oblivious to our rules and under no obligation to reveal its complexities or distribute its wealth. We should be careful stewards of our own home, and cautious, humble visitors to others.

When I look at the night sky through my telescope, I see the firmament not as it is now, but as it was long ago. By studying Jupiter's moons from the observatory in Uppsala, I know that light travels at fantastic but finite speed, so the images of faraway objects take thousands or millions of years to reach my eye. I can see stars that may no longer exist in chronological time. Similarly, the weather records that Professor Burman and I started to compile in 1722, and the different scales we used to monitor temperature, preserve the reality of past events so that they can be interpreted much later. To understand our past and present, and improve our future we must therefore first comprehend the nature of time itself. It is never static and never linear. It is the woven fabric of existence in which we are clothed, which we can neither alter nor remove.

My work has also taught me that the greatest phenomena and most irresistible forces are often simply the aggregate of countless, smaller actions and reactions compounded repeatedly over time and in near-infinite diversity. The magnetic fields that create the beautiful displays of the aurora borealis, or the land and sea movements I have investigated in the Baltic, are both the products of interdependent dynamics, which our senses cannot detect in the moment. But these energies exist nonetheless and alter the world forever. We would do well to apply these principles of unity and synergy to our own mortal endeavours.

The special human gifts of which I spoke earlier carry with them particular responsibilities, privileges and possibilities. I contend that each man and woman has a duty to apply their intellect and rationalism to kind, positive and practical purposes, and in turn to encourage and educate others to do the same. The process of discovery has no end, and we must constantly remember that there is much we do not yet know, comprehend or even perceive. Hence, we should pursue questions as much as answers.

Research by my dear friend Linnaeus suggests that the diverse inhabitants of this planet are not fixed, but rather that they continually appear, adapt and disappear. It follows then that, if our own time here is limited, we can — if we choose — contemplate and prepare for our demise. Now that my own constitution is weak, I find myself considering death and the traces I shall leave behind. People who are fortunate to live to an old age also do this quite naturally as individuals. So why should we not do the same collectively?

I wish that humanity's curiosity, discovery and science can in the future be focused upon fulfilling our brief destiny, so that we may pass on this wonderful Earth alive and untrammelled to those that follow us. Acknowledging and accommodating our ultimate end could be the key to our best possible future.

And Celsius

Uppsala, April 1744

BIBLIOGRAPHY

Selected reading, titles and other sources used to research this book, which are not referenced in the notes (alphabetically by author):

Aldersley-Williams, H., *Tide: The Science and Lore of the Greatest Force on Earth* (Penguin, 2017).
Barrie, D., *Sextant: A Voyage Guided by the Stars and the Men Who Mapped the World's Oceans* (William Collins, 2015).
Beckman, O., *Anders Celsius* (Uppsala Universitet, 2003).
Beckman, O., 'Anders Celsius and the Fixed Points of the Celsius Scale', *European Journal of Physics* (1997).
Bergström, H., 'The Early Climatological Records of Uppsala', *Geografiska Annaler*. Series A, Physical Geography (Taylor & Francis Ltd, 1990).
Crane, N., *Latitude: The Astonishing Adventure That Shaped the World* (Penguin, 2019).
Crane, N., *Mercator: The Man Who Mapped the Planet* (Phoenix, 2003).
Ekman, M., *A Geophysical and Astronomical Analysis of an Old Painting of the Stockholm Sluice* (Summer Institute for Historical Geophysics, 2011).
Ekman, M., *An Investigation of Celsius' Pioneering Determination of the Fennoscandian Land Uplift Rate, and of His Mean Sea Level Mark* (Summer Institute for Historical Geophysics, 2013).
Ekman, M., *Calculation of Historical Shore Levels Back to 500 A.D. in the Baltic Sea Area Due to Postglacial Rebound* (Summer Institute for Historical Geophysics, 2017).
Ekman, M., *The Man Behind 'Degrees Celsius' A Pioneer in Investigating the Earth and its Changes* (Summer Institute for Historical Geophysics, 2016). Available from www.historicalgeophysics.ax.
Ekman, M., & Ågren, J., *A Partial Reanalysis of the French Arc Measurement at the Arctic Circle to Prove Newton's Theories* (Summer Institute for Historical Geophysics, 2013).
Ekman, M., & Ågren, J., *A Study of Celsius' Astronomical Latitude Determination of the Uppsala Observatory Using Satellite Positioning and Deflections of the Vertical* (Summer Institute for Historical Geophysics, 2012).
Franzon, A., *A Guide to Uppsala Cathedral* (Church of Sweden, 2016).
Härkönen, L., *Tornio: Moments in the Flow of Time* (Innoplus, 2021).
Harnesk, H., *Linnaeus: Genius of Uppsala* (Uppsala Universitet, 2020).
Hempel, S., *The Medical Detective: John Snow, Cholera and the Mystery of the Broad Street Pump* (Granta, 2006).

Liiffe, R., *Aplatisseur du Monde et de Cassini: Maupertuis, Precision Measurement, and the Shape of the Earth in the 1730s* (Institute of Historical Research, London University, 1993).

Ince, R., *The Importance of Being Interested: Adventures in Scientific Curiosity* (Atlantic Books, 2022).

Johnson, S., *The Invention of Air: An Experiment, a Journey, a New Country, and the Amazing Force of Scientific Discovery* (Penguin, 2009).

King, D., *Finding Atlantis: A True Story of Genius, Madness, and an Extraordinary Quest for a Lost World* (Harmony Books, 2005).

Lin, Y.S., *Fahrenheit, Celsius and Their Temperature Scales* (Powerkids Press, 2012).

Littmarck, T., *Gamla Uppsala: From Ancient to Modern Time* (Old Uppsala Parish Vestry, 2002).

Mann, M.E., *The New Climate War: The Fight to Take Back Our Planet* (Scribe Publications, 2021).

Nordenmark, N.V.E., *Anders Celsius: Professor i Uppsala 1701–1744* (1932).

Redelius, G., *Kristina Kyrka i Sala* (1987).

Robinson, K.S., *The Ministry for the Future* (Orbit, 2020).

Rovelli, C., *Seven Brief Lessons on Physics* (Penguin, 2016).

Sobel, D., *Longitude: The True Story of a Lone Genius Who Solved the Greatest Scientific Problem of His Time* (Walker and Company, 2005).

Solarić, M. & Solarić, N., *The French Geodetic and Scientific Expedition to Lapland* (University of Zagreb, 2014).

Spencer, N., *Magisteria: The Entangled Histories of Science and Religion* (Oneworld, 2023).

Steffen, H., *Glacial Isostatic Adjustment: An Introduction* (EGSIEM Summer School Potsdam, 2017).

Stempels, H.C., *Anders Celsius' Contributions to Meridian Arc Measurements and the Establishment of an Astronomical Observatory in Uppsala* (Uppsala University, 2011).

Stünkel, K.M., *Meeting at the Philosopher's Stone: The Encounter of Enlightenment and Indigenous Religion in Maupertuis' Expedition to Lapland (1736–1737)* (Ruhr-Universität Bochum, Germany, 2020).

Sundqvist, B., *Anders Celsius ock Skådetornet* (Uppsala Universitet, 2022).

Svard, B., *Silverbrytning, Búder och Tankar* (2013).

Todhunter, I., *A History of the Mathematical Theories of Attraction and the Figure of the Earth* (Cambridge University Press, 2015).

Vincent, J., *Beyond Measure: The Hidden History of Measurement* (Faber & Faber, 2022).

Widmalm, S., 'Auroral Research and the Character of Astronomy in Enlightenment Sweden', *Acta Borealia, A Nordic Journal of Circumpolar Societies* (2012).

ACKNOWLEDGEMENTS

Researching and completing this book was, like most of Celsius' work, a collaborative undertaking. Thanks to the generosity and help of many dozens of people, it was also hugely enjoyable and rewarding.

In Sweden, it was my friend Ralph Edwards' chance pointing out of the original observatory building in the centre of Uppsala that first brought Anders Celsius to my attention. Then it was Martin Ekman, with his unrivalled knowledge of my subject's life and passion for his science, who inspired me to tell the story. Sarah Covington at City University of New York also provided welcome early encouragement for the project.

Through research trips to Sweden, Finland and the Netherlands I have been fortunate to meet and learn from many kind and gifted individuals. Without exception, they have shared their knowledge and resources in a spirit of open, international assistance, which stands as its own tribute to Celsius' freedom of thought across nations and cultures.

In Uppsala, Eric Stempels, Tom Stevens, Christoffer Hallgren, Anna-Zara Lindbom, Rebecca Flodin, Kia Hedell, Moa Bergkvist, Christel Kraft, Lena Urtel, Jörgen Runeby, Petra Stärkman, Karl Lagergren, Annika Franzon and Hans Bergström all lent invaluable support and allowed me to see some of Celsius' most significant places, documents and artefacts.

Elsewhere in Sweden, Niklas Ulfvebrand (at Sala Silver Mine) and Jens Lidberg (whose powerful boat took me out to Lövgrund Island), provided some of the most thrilling and memorable experiences of following in Celsius' footsteps. Olov Amelin from the Jamtli Museum in Östersund also gave helpful advice and contacts.

My visit to the Arctic Circle to retrace the route of Celsius' 1736–37 expedition was part funded by a research grant from the Scientific Instrument Society. And it was brought to life by Veli-Markku and

Tuomo Korteniemi of the De Maupertuis Foundation, Jarno Niskala, Ilkka Halmkrona, Riitta Yrjanheikki, Eero Ylitalo, Miia Kallioinen, Janne Tolvanen, Sisko Kalliokoski, Lauri Posio, Joni-Pekka Karjalainen and the late Osmo Pekonen.

In the UK, I'm grateful to Cherry Mosteshar at the Oxford Editors and Richard Milbank at Head of Zeus, who assured me that the idea of attaching a biography of Anders Celsius to a polemic on aspects of modern climate change 'had legs'. My mentor Rebecca Abrams then guided me in how to structure the book, clarify its message and find its voice.

From the Oxford Centre for Life Writing, the prime members and friends who assisted in turning the idea into reality were Alice Little, Kate Kennedy, Hermione Lee, Elleke Boehmer, Hana Navratilova, Kate Elliott, Rebecca Gowers, Charles Pidgeon, Charlie Lee-Potter, Tamarin Norwood, Helen de Borchgrave, Richard Walker, Penelope Gardner-Chloros, Mary Black, Victoria Phillips, Shirley Tung, Isabelle Meuret, Carol Peaker, Oge Nwosu, Lizzie Wingfield, Frances Larson and Michael Meyer.

At Wolfson College and the University of Oxford, Tim Hitchens, Victoria Forster, Karen Konopka, Fiona Wilkes, Julie Howard, Tarje Nissen-Meyer, Moritz Riede, Diane Mackay, Tom Nelson, Noi Loket, Ola Bunning, Janice Tirda, Steve Lewis, Cindy Pascal, Liz Baird, Huw David, Peter Stewart, Michael Godfrey and Darek Sobon all helped me to make the most of my visiting scholarship.

The members of GROW (Group of Oxford Writers: Robert Bullard, Pauline Cakebread, Richard Cullen, Nigel Moor, Andrew Bartholomew and Elle Montoya) and Didcot Writers gave friendly and honest feedback throughout the writing process. This was especially useful during the wholly online days of pandemic lockdowns. Duncan Laing provided detailed editing advice on the manuscript.

Stephen Burt from the Department of Meteorology at the University of Reading and Nicolàs de Hilster from the Scientific Instrument Society supported my research with stimulating visits and the opportunity to publish blogs, plus essential fact-checking, proofreading and photographs, while Jen Farquharson and Joseph Holloway facilitated a series of articles for the British Society for the History of Science. The expert knowledge of Jonathan Bushell, Rupert Baker and Ellen Embleton at the Royal Society in London gave me hands-on access to some of Celsius' and his contemporaries' writings. Thanks also to Johan Rockström and Claudia Köhler for use of their material from the 2022 Celsius tricentennial lectures.

Since most of Celsius' work was in either Latin or Swedish, I am grateful to Martin Ekman, Ludovico Oddi and Marie Lindquist for their help in translating letters, journals and publications into English. The spellings of some historical Swedish words and names vary, but I have tried to be consistent by using modern, Anglicised versions.

In the later stages of finishing and preparing the book for publication, my editor Claire Hartley and her colleagues at The History Press were a tremendous source of guidance and practical support. Melanie Gee mastered the mass of historical and scientific material and cross references to index the copy and images in a way that I hope readers will find helpful. Most of the historical images included here are public domain by virtue of their age, but I have credited other picture sources and licences where appropriate.

Deep gratitude goes to my 'ideal readers' and special friends Robert Smith, Mark Hammond, Richard Sewell, Bruce Hugman and Andrew Baum. And to Nicola Winn and all my colleagues at Creative Bridge for giving me the time and flexibility to complete the book.

Biographers must necessarily live with their subjects for prolonged periods, and the families of biographers are destined to live with whatever this entails. My immeasurable thanks to my wife Pippa and sons Robert and Frazer for their never-failing belief in the project and loving help, companionship and motivation to make it happen.

Ian Hembrow
Oxford
Summer 2024

NOTES

1 carlleonarddotcom.wordpress.com/2012/10/15/we-live-submerged-at-the-bottom-of-an-ocean-of-air-evangelista-torricelli-2/.
2 Uppsala Celsius Lecture, 2023.
3 Narasu, P.L., *The Essence of Buddhism* (Srinivasa Varadachari & Co., 1907).

Prologue

1 www.un.org/sustainabledevelopment/climate-change/#:~:text=To%20limit%20warming%20to%201.5,to%20net%20zero%20by%202050.
2 United Nations Conference on Biodiversity, 2010. See www.cbd.int/cop10.
3 de Menocal P., *Nature*, obituary, 26 March 2019.

Chapter 1: Risen from the Sea

1 Rudbeck, O. *Atlantica: Atland eller Manheim*, 1675–99.
2 Celsius, A., 'Observations of the Aurora Borealis Made in England', *Philosophical Transactions of the Royal Society of London*, 1735.
3 Heyman, H.J., sok.riksarkivet.se/Sbl/Presentation.aspx?id=14758.
4 Ferner, B., *Resa i Europa*, 1758–62.

Chapter 2: Forged in Flames

1 Eenberg, Johan, *En utförlig relation om den gruvliga eldsvåda*.
2 www.svenskakyrkan.se/uppsala/building.
3 www.gustavianum.uu.se/about-us/.
4 King, D., *Finding Atlantis* (2005).

Chapter 3: A City of Learning

1 www.uu.se/en/about-uu/history/summary/.
2 www.thereformation.info/scandinavia/.
3 www.lutheran.ro/iratok/ca_en.pdf.

Chapter 4: A Family of Ambition

1. Dahlin, E.M., *Contributions to the History of Mathematics in Sweden Before 1679*, PhD thesis, 1875.
2. alumni.blogg.lu.se/a-cell-for-the-noisiest-and-most-unruly-or-where-was-the-university-lock-up/.
3. afuu.org/digitalAssets/141/a_141715-f_cracking-the-runic-code.pdf.
4. Celsius, O., *Hierobotanicon Sive De Plantis Sacrae Scripturae*, Amsterdam, 1748.
5. King, D., *Finding Atlantis*.

Chapter 5: A Frustrated Father

1. urn.kb.se/resolve?urn=urn:nbn:se:alvin:portal:record-253117.
2. Copernicus, Nicolaus, *De Revolutionibus Orbium Coelestium*, Nuremberg, 1543.
3. Descartes, René, *Discours de la Méthode Pour bien conduire sa raison*, 1637.
4. anders.chydenius.fi/en/life/sweden-in-the-age-of-freedom/.
5. www.rocketstem.org/2020/12/12/ice-and-stone-comet-of-week-51/.
6. Bilberg, J., *A Voyage of the Late King of Sweden*, 1698.
7. Rudbeck, O., *Filii Nora samolad sive Lapponia illustrata*, 1701.

Chapter 6: The Age of Enlightenment

1. plato.stanford.edu/entries/newton-principia/.
2. www.gutenberg.org/files/10615/10615-h/10615-h.htm.
3. Von Höpken, A.J., *Åminnelse-tal öfver astronomiæ professoren*, 1745.
4. Rudbeck, O., *Filii Nora samolad sive Lapponia illustrata*.
5. Newton, I., *Opticks*, 1704.
6. Sébastien le Clerc, *Nova Geometria Practica Super Charta et Solo*, 1692.
7. www.auroraorchestra.com/2019/05/28/pythagoras-the-music-of-the-spheres/.
8. Nicomachus of Gerasa, *Encheiridion Harmonikes* (*Manual of Harmonics*), c. 100 BCE.

Chapter 7: The Young Professor

1. Torricelli, E., Letter to Michelangelo Ricci, Rome, 1644. See www.lindahall.org/about/news/scientist-of-the-day/michelangelo-ricci.
2. Linnaeus, C., quoted in Kjellin, M.E.N., *Genuine Falun Red* (Prisma, 1999).
3. Celsius, A., *Experimentum, in Argenti-Fodina Salana*, 1724.
4. Celsius, A., *Arithmetica Ellet Raekne'konst*, 1728.
5. See uu.diva-portal.org/smash/get/diva2:1567296/FULLTEXT01.pdf.
6. Linnaeus, C., *Praeludia Sponsaliorum Plantarum*, 1729.
7. www.linnean.org/learning/who-was-linnaeus/career-and-legacy.
8. Von Linné, C., *Lachesis Lapponica* (*A Tour in Lapland*), English translation by J.E. Smith, 1811.

Chapter 8: Celsius' Grand Tour

1. See map iii.

Chapter 9: Two Countries, Four Sisters

1 Helly, O.D., and Reverby, S., *Gendered Domains: Rethinking Public and Private in Women's History* (Cornell University Press, 1992).
2 Petropolitanae Academia Scientarum, *Musei Imperialis Petropolitani*, Part 3, St Petersburg, 1703.
3 Copernicus, N., *De revolutionibus orbium coelestium*, Nuremberg, 1543.
4 site.unibo.it/accademiascienzebologna/en/members/class-of-physical-sciences.
5 Manfredi, E., *Ephemerides motuum Coelestium* (*Ephemerides of Celestial Motion*), Bologna, 1715.
6 Barnardi, G., *The Unforgotten Sisters: Female Astronomers and Scientists before Caroline Herschel* (Springer Praxis, 2016).

Chapter 10: In Paris

1 MacDonogh, G., *Frederick the Great: A Life in Deed and Letters* (St Martin's Griffin, 2001).
2 Newton, I., *Philosophiae Naturalis Principia Mathematica* (*Mathematical Principles of Natural Philosophy*), Cambridge, 1687.
3 Cassini, J., *De la grandeur et de la figure de la terre*, 1720.
4 De Maupertuis, P.L.M., *Discourse on the Shapes of the Heavenly Bodies*, 1732.
5 Quoted in Beeson, D., *Maupertuis: An Intellectual Biography* (Voltaire Foundation, 1992).
6 Voltaire, *Histoire du Docteur Akakia et du Natif de St Malo*, 1752.
7 Available at www.academie-sciences.fr/fr/Transmettre-les-connaissances/histoires-de-l-academie-royale-des-sciences-memoires-et-proces-verbaux-des-seances-numerises-par-la-bibliotheque-nationale-de-france.html.
8 Linnaeus, C., *Flora Lapponica*, Amsterdam, 1737.

Chapter 11: In London

1 Sir Hans Sloane's diary, quoted in Bakan, A., *Ideology and Class Conflict in Jamaica: The Politics of Rebellion* (McGill-Queen's University Press, 1990).
2 Sloane, H., *A Voyage to the Islands of Madera, Barbados, Nieves, Saint Christophers and Jamaica 1707–1725*.
3 Quoted in Delbourgo, J., *Collecting the World: The Life and Curiosity of Hans Sloane* (Allen Lane, 2017).
4 Ortolja-Baird, A., *'Chaos naturae et artis': Imitation, Innovation, and Improvisation in the Library of Sir Hans Sloane* (Edinburgh University Press, 2020).
5 *Observations on the lunar eclipse of March 15, 1736 made at Mr Graham's house in Fleet Street by Mr Celsius, FRS. with a reflecting telescope of eleven inches, magnifying sixty three times and made at Edinburgh*, The Royal Society.
6 Old Turkish proverb to describe the intensity of the country's coffee.
7 See royalsociety.org/about-us/history/.

Chapter 12: Towards the Pole

1 Outhier, Réginald, *Journal d'un voyage au nord, en 1736. & 1737*, 1744.
2 *Ibid.*
3 *Ibid.*
4 *Ibid.*

5   Ibid.
6   Ibid.
7   De Maupertuis, P.L.M., et al., *The Figure of the Earth*, 1738.
8   Outhier, Réginald, *Journal d'un voyage au nord*.
9   Ibid.
10 Ibid.
11 Ibid.
12 Ibid.
13 Ibid.
14 Ibid.

Chapter 13: Arctic Summer

1   Outhier, Réginald, *Journal d'un voyage au nord*.
2   Terrall, M., *The Man Who Flattened the Earth* (University of Chicago Press, 2002).
3   De Maupertuis, P.L.M., et al., *The Figure of the Earth*.
4   Ibid.
5   Celsius, A., Letter to Sir Hans Sloane, 4 December 1736. Library of the Royal Society, London.
6   De Maupertuis, P.L.M., et al., *The Figure of the Earth*.

Chapter 14: Arctic Winter

1   Outhier, Réginald, *Journal d'un voyage au nord*, p.74.
2   Anonymous, *Anecdotes physiques et morales*, Paris, 1738 (*party games such as blind man's buff).
3   Celsius, A., *De observationibus pro figura Telluris determinanda*, 1738. (Summary in *Philosophical Transactions of the Royal Society of London*, 41, pp.371–82.)
4   De Maupertuis, P.L.M., 'Relation d'un voyage au fond de la Laponie' in *Oeuvres*, 1756.
5   De Maupertuis, P.L.M., *Oeuvres*, Vols I–IV, 1756.
6   Knut, M.S., *Meeting at the Philosopher's Stone: The Encounter of Enlightenment and Indigenous Religion in Maupertuis' Expedition to Lapland (1736–37)* (Center for Religious Studies, Ruhr-Universität Bochum, 2020).

Chapter 15: Fighting for the Truth

1   De Maupertuis, P.L.M., et al., *The Figure of the Earth*.
2   Terrall, M., 'Representing the Earth's Shape: The Polemics Surrounding Maupertuis's Expedition to Lapland', *Isis*, Vol. 83, No. 2, 1992.
3   Clairaut, A.C., Library of the Royal Society, London.
4   Terrall, M., 'Representing the Earth's Shape: The Polemics Surrounding Maupertuis's Expedition to Lapland', *Isis*, Vol. 83, No. 2.
5   Ibid.
6   Ibid.
7   Ibid.
8   Ibid.
9   Celsius, A., *De observationibus pro figura Telluris determinanda*.
10 Ibid.
11 Terrall, M., *The Man Who Flattened the Earth*.

12 Ibid.
13 Cassini, J., *Réponse à la dissertation de M. Celsius*, Paris, 1738.
14 Ferreiro, L.D., *Measure of the Earth: The Enlightenment Expedition That Reshaped Our World* (Basic Books, 2011).
15 Crane, N., *Latitude*.
16 Terrall, M., *The Man Who Flattened the Earth*.
17 Ibid.
18 Ibid.
19 Voltaire, F., 'Quatrième discours de la modération en tout' in *Oeuvres*, 1738/1752.
20 De Maupertuis, P.L.M., et al., *The Figure of the Earth*.
21 Terrall, M., *The Man Who Flattened the Earth*.
22 Ibid.
23 Ibid.
24 Ibid.
25 Ibid.
26 De Maupertuis, P.L.M., *Examen désintéressé des différentes ouvrages*, Oldenbourg, 1738 [1740].
27 Terrall, M., *The Man Who Flattened the Earth*.
28 De Maupertuis, P.L.M., *Examen désintéressé des différentes ouvrages*.
29 Cassini de Thury, C.F., *Sur les opérations géometriques*, Paris, 1740.
30 Ibid.
31 Ibid.
32 Terrall, M., *The Man Who Flattened the Earth*.
33 La Condamine, C.M., *Mesure des trois premier degrés du méridien dans l'hémisphère austral*, Paris, 1751.
34 Celsius, A., Library of the Royal Society, London.

Chapter 16: Serving and Observing

1 Celsius, A., *Anteckningar angående norrsken och andra meteorologiska fenomen*, Carolina Rediviva Library, University of Uppsala.
2 Celsius, A., *Nyttan af et astronomiskt observatorium uti Swerige*, Uppsala, 1739.
3 Celsius, A., *Anteckningar angående norrsken och andra meteorologiska fenomen*.
4 Ibid.
5 Ibid.
6 See www.iau.org/public/themes/constellations/.

Chapter 17: A Vast, Profound Truth

1 Ekman, M., *The Changing Level of the Baltic Sea During 300 Years: A Clue to Understanding the Earth* (Summer Institute for Historical Geophysics, 2009).
2 Ekman, M., *The Changing Level of the Baltic Sea During 300 Years*.
3 Roden, G.I., and Rossby, H.T., 'Early Swedish Contribution to Oceanography: Nils Gissler (1715–71) and the Inverted Barometer Effect', *Bulletin of the American Meteorological Society*, Vol. 80, 1999.
4 Celsius, A., *Anmärkning om vattnets förminskande så I Östersiön som Vesterhafvet*, Kongl. Swenska Wetenskaps Academiens Handlingar, 1743.
5 Roden, G.I., and Rossby, H.T., 'Early Swedish Contribution to Oceanography'.

6   Ibid.
7   Dalin, O., *Svea Rikes Historia*, Vols 1–4, 1747–62.
8   Roden, G.I., and Rossby, H.T., 'Early Swedish Contribution to Oceanography'.
9   Hammarklint, T., *Swedish Sea Level Series: A Climate Indicator* (Swedish Meteorological and Hydrological Institute, 2009).
10 Venetz, I., *Mémoire sur les Variations de la température dans les Alpes de la Suisse*, 1821.
11 Agassiz, L., *Études sur les glaciers*, 1840.
12 Ganopolski, A., Winkelmann, R., and Schellnhuber, H.J., 'Critical Insolation – CO2 Relation for Diagnosing Past and Future Glacial Inception', *Nature*, 2016.

Chapter 18: The Infinite and the Invisible

1   See scientificinstrumentsociety.org.
2   Hiorter, O., *Om magnet-nålens åtskillige ändringar*, Uppsala, 1747.
3   A micron (μm or micrometre) is 1 millionth of a metre. So 1,000 microns (μm) = 1 millimetre (mm).

Chapter 19: One Hundred Steps

1   Halley, E., *An Account of Several Experiments Made to Examine the Nature of Expansion and Contraction of Fluids by Heat and Cold*, The Royal Society, 1693.
2   Frängsmyr, T., Heilbron, J.L., and Rider, R.E. (eds), *The Quantifying Spirit in the 18th Century* (University of California Press, 1990).
3   Chang, H., *Inventing Temperature: Measurement and Scientific Progress* (Oxford Academic, 2004).
4   Beckman, O., and Moberg, A., 'The Celsius Thermometer', *Weather*, Vol. 55, 2000.
5   Celsius, A., *Observationer om twänne beständiga Grader på en Thermometer* (*Observations on two constant degrees on a thermometer*), Stockholm, 1742.
6   See www.bipm.org/en/home.
7   See tobefrank.se/in-english/.

Chapter 20: Death of a Star

1   Von Höpken, A.J., *Åminnelse-tal Öfver astronomiæ professoren och Kongl. vetenskaps academiens medlem herr Anders Celsius*, Stockholm, 1745.
2   Hiorter, O., *Underdånödmiukt Memorial*, Uppsala, 1745.
3   Halley, E., *An Account of Several Experiments Made to Examine the Nature of Expansion and Contraction of Fluids by Heat and Cold*.

Chapter 21: Noble Successors

1   See mathworld.wolfram.com/Pi.html.
2   See mathworld.wolfram.com/GoldenRatio.html.
3   Letter from Wargentin to Hiorter, 8 March 1750, Uppsala University Library.
4   Hiorter, O., Deed of gift to Uppsala University, 29 July 1747.
5   Malthus, T., *Essay on the Principle of Population* (London, 1798).
6   Darwin, C., *On the Origin of Species* (John Murray, 1859).

7   Blunt, W., *Linnaeus: the Compleat Naturalist* (Frances Lincoln Publishers Ltd, 2001).
8   Linnaeus, C., *Flora Lapponica*.
9   Linnaeus, C., *Systema Naturae*, 1735.
10 Linnaeus, C., *Philosophia Botanica*, 1751.
11 Linnaeus, C., *Species Plantarum*, 1753.

Chapter 22: A Safe and Just Earth for All

1   See www.permaculture.org.uk/knowledge-base/basics.
2   See www.quantamagazine.org/how-will-the-universe-end-20230222/.
3   See www.ipcc.ch/report/sixth-assessment-report-working-group-ii/.
4   Durkheim, E., *Les formes élémentaires de la vie religieuse* (*Elementary Forms of Religious Life*), translated, George Allen & Unwin Ltd, 1912.
5   See www.quantamagazine.org/how-will-the-universe-end-20230222/.
6   See www.europarl.europa.eu/news/en/headlines/economy/20151201STO05603/circular-economy-definition-importance-and-benefits.
7   See www.amsterdam.nl/en/policy/sustainability/circular-economy.
8   See www.2000-watt-society.org/what.
9   See www.fao.org/sustainability/news/detail/en/c/1274219.
10 See www.labiotech.eu/in-depth/precision-fermentation-food-supply/.

Epilogue: Anders Celsius' Message to the Twenty-First Century

1   See Chapter 9.
2   See readwj.files.wordpress.com/2022/05/a-message-to-the-twenty-first-century-isaiah-berlin.pdf

INDEX

Note: AC = Anders Celsius. *Italicised* page references denote illustrations. Surnames containing the separate prefix 'de' are entered under the element following the prefix, e.g. Maupertuis, Pierre Louis Moreau de.

Aavaksaksa, 169, 177
l'Académie des Sciences, France, 124–5, 129, 181, 187–9, *189*, 190, 197, 201
Adolf Frederick, Crown Prince of Sweden, 251–2
Agassiz, Louis, 227
air pressure *see* barometers and air pressure measurement
Algarotti, Francesco, 119–21, *120*, 145, 199
almanacs, 54, 64, 82, 109, *110*, 212, 256
Ångströmlaboratoriet, Uppsala University, 213–4, 254
Arctic Circle, *20*, 161, *173*, 177
Arctic expedition (1736–1737)
    Abbé Outhier's accounts, 149–50, 152, 154, 156, 157, 158, 159, 166
    consorting with women, 180, 193–4
    Dunkerque to Stockholm, 150–6, *153*
    latitude distance measurement
        baseline and measuring points, 163–7, 168–72, *169*, *170*, *173*, *176*, 177–8, *179*
        baseline measurement, 178–9, *179*
        calculations, 180–1
        debates about findings, 185, 187–90, 200–1, 202
        instruments and apparatus, 152, 172–4, *174*, 178, 187, 214
        measurement inaccuracies, 201–2
        observatories and astronomical observations, 174, 175–6, *176*, 181
        publication of findings, 189–91, 195, *195*, 199–200, *200*, 202–3
    Maupertuis monument, Pello, 161–3, *162*, 183
    party, 131–2, 149, 167
    preparations, 132, 133–4, 138, 145–6, 149–50
    Le Prudent, 150–4, *151*, 155, 156, 159, 163, 184
    return trip, 184
    Stockholm to Tornio, 157–9, *160*
    Stone of Käymäjärvi, 181–3, *182*
    Tornio, 163–4, 166–7, 171, 172–4, 175–6, 180–1
astronomy
    AC's education in, 74–5, *75*
    AC's Grand Tour encounters, 107–9, 111–3, 117–8, 122–4, 131–2, 138–43
    in AC's heredity, 52, 53, 54, 55–64, *58*
    AC's Moon thesis, 89
    almanacs, 54, 64, 82, 109, *110*, 212, 256
    and the Church, 47, 58–9, *58*
    comets, 59, *60*, 108, 240, 251–2, *252*, 253, 256
    instruments, 109, 132, 138, 145–6, 152, 164, 172–4, *174*, 184, 187, 208, 213–4, 231–2
    Jupiter and moons, 108, 141, 260–1
    latitude distance measurement *see under* Arctic expedition (1736–1737)
    Northern Lights, 32, 108, 109–11, *110*, 129, 141, 152, 234–5

observatories *see* observatories
professorship, Uppsala University, 49, 52, 57, 63, 75, 253
    AC's professorship, 91, 121, 132, 207, 208–10
aurora borealis *see* Northern Lights

barometers and air pressure measurement, 81–4, 87–8, 222
    see also weather observations
Basilica di San Petronio, Bologna, 115
Benzelius, Eric, *190*, 252
    AC's letters to, 90, 107, 134, 138, 189, 208, 252
Berlin, *22*, 107–9, *107*
Berlin, Sir Isaiah, 281–2
Bernard, François, 150, 154, 163
Bernoulli, Johan, 128, 186, 201
Biaudos, Charles-Louis de, 155, 156
Bilberg, Johannes, 60–1, 68–9, *69*, 95
binomial nomenclature, 264, *266*–7
Biurman, Georg, 103
    see also Grand Tour, of AC (1732–1736)
Bologna, *22*, 111–6, *114*, 281
botany, 92–8, *94*, *96*, *97*, 136–7, 224, 225, 264–5
Bouguer, Pierre, 129–30, 192
    see also Peru expedition (Louis Godin's party)
Bradley, James, 138, 140–1, *140*, 189, 195
Browallius, Johannes, 224–5, *224*
Burman, Eric, 68, 75, 76–8, *76*, 89, 90–1, *273*

calculus, 70, 72–3
Cambridge, 141
Camus, Charles Étienne Louis, 131, 184, 197
    see also Arctic expedition (1736–1737)
Carolina Rediviva Library, Uppsala, 73
Cassini, Giovanni Domenico (Jean-Dominique), 125, *125*
Cassini, Jacques, 125, 126, 187–90, 194, 197, 199, 200
Cassini de Thury, César-François, 201, *202*
Celsia, Sara-Märta, 37, 64, 67, 209
Celsius, Anders, 23–4, *284*
    air pressure experiments, 82–4, 87–8
    appearance, 12, 208, 257
    Arctic expedition *see* Arctic expedition (1736–1737)
    astronomy
        builds Uppsala observatories, 208, 210–3
        Grand Tour encounters, 107–9, 111–3, 117–8, 122–4, 131–2, 138–43

        latitude distance measurement *see under* Arctic expedition (1736–1737)
        Moon thesis, 89
        Northern Lights records analysis, 109–11, *110*
        observes Great Comet (1744), 251–2
    career
        astronomy professorship, 91, 121, 132, 207, 208–10
        Fellowship of Royal Society, London, 142–3, *144*
        pension from Louis XV, 184–5, 190, 254
        rectorship, 252
        secretary to Swedish Royal Society of Sciences, 208–9
        teaching at King's Barn, 89, *90*
        unpaid university appointments, 89, 90
    character, 33–4, 105, 190, 258–9, 276, 278
    death, 252, 253
        commemorative address (1743), 68
        funeral and memorial tribute, 68, 253, *258*
        inventory of belongings, 254, 278
        tombstone and memorial plaque, 27, *28*, 265
    education, 68–71, 73–5, *74*
    Grand Tour *see* Grand Tour, of AC (1732–1736)
    gravity measurements, 239–40
    illness, 203, 236, 252–3
    imagined message to present day times, 281–4
    infancy and childhood, 37, 64
    language skills, 91, 105, 107, 140
    legacy and relevance today, 35, 274–6
    library, 214, 254, 259
    long-term vision, 32, 35, 274
    magnetic field measurements, 232–4, *233*, *234*
    portrait, 208, 214, 250, *284*
    publications, 88, 89, *90*, 111, 189–90, 202–3, 212–3
    quoted
        on air pressure experiments, 88
        Grand Tour journal, 109, 207
        letter to Carl Gyllengborg, 210
        letter to Cromwell Mortimer, 203
        letter to Hans Sloane, 171
        letter to Joseph-Nicolas Delisle, 190
        letter to Magnus Beronius, 117–8
        letters to Eric Benzelius, 90, 107, 134, 138, 189, 208, 252
        letters to his mother, 113, 132

on life after death, 253
on magnetic field measurements, 234
on observational methods, 32
on philosophy, 90
on sea level changes, 222–3
on shape of Earth, 190
on Uppsala weather records, 246
sea level investigations and theories, 216, 219–21, 222–3, 225, 230
statue, 12, 240, 256
temperature scale, 241, *241*, 245–7, 248–9
weather measurement, 78, 81, *273*
Celsius, Johan, 53–4
Celsius, Magnus, 51–3, 54, 92, 181
Celsius, Nils, 37–9, 53, 57–60, *58*, 63, 64, 66–7, 69, 74, *75*, 88–9, 213
Celsius, Olaf (the Elder), 54, 92–3, *92*, 95
Celsius, Olaf (the Younger), 265
Celsius family tree, *19*
Charles IX, King of Sweden, *48*, 49
Charles XI, King of Sweden, 56, 59, 60–1, *62*
circular economy, 278–9, *279*
Clairaut, Alexis Claude, 131, 197
Clement XII (Pope), 117–8
climate change
    ice ages, 226–7, 229–30, 277
    man-made, 34, 230, 275, 277
    Pehr Wargentin's observations, 261–2
coffee houses, London, 138–9
Colbert, Jean-Baptiste, 124
comets, 59, *60*, 108, 240, 251–2, *252*, 253, 256
consumption (tuberculosis), 236, *236*, 252–3
Copenhagen University, 46, 47
Copernicus, Nicolaus, 111, *112*

Dalin, Olof, 223–4, *224*
Davia, Cardinal Gianantonio, 117–8
De Maupertuis Foundation, 167
    see also Maupertuis, Pierre Louis Moreau de
Delisle, Joseph-Nicolas and family, 122, 190, 243–5, 246
Du Rietz, Lieutenant Colonel, 167
Duhre, Anders Gabriel, 71–3, *72*
Dunkerque, *22*, 149–50, *151*, *153*
Durkheim, Émile, 276–7

Earth, properties of
    distance from Sun and planets, 261
    gravitational strength, 239
    magnetism, 232–6
    shape, 126–7, 129–30, 164, 172, 180–1, 185, 186–91, 192–3, 195–7, *195*, *198*, 199–202, *200*, 239

size, 125–6
structure and composition, 227, 228–9, *229*, 235
Ekman, Martin, 220
Ekström, Daniel, 213, *233*, 246
Elvius, Pehr (the Younger), 256, 260
Elvius, Petrus, 63, 64, 68, 71
Enlightenment, 65–6, 245, 275
equatorial expedition (Louis Godin's party), 129–30, 146, 159, 191–3, *191*, *192*, *193*
Esmark, Jens, 226

Fahrenheit, Daniel Gabriel (and temperature scale), 34, 242–3, *244*, 249
Falun copper mine, *20*, 87–8
Ferdinand II, Grand Duke of Tuscany, 241, *244*
Ferner, Bengt, 34, 70–1
Figrelius, Sara, 53, *53*, 54
food and kitchens, 18th century, 67, *68*
fossils, 226, 227, 228
Frederic I, King of Sweden, 155, 156, *156*, 184
Frederick the Great, King of Prussia, 120–1

Galen (Greek/Roman philosopher), 240, *244*
Galileo Galilei, 81, 231, 235, 241, *244*, 260–1
Gamla Uppsala, 27, *28*, 29–31, *30*, *31*, 32–3, 54, 158, 253
geometry, 68, 70, 73–4, *74*
Gissler, Nils, 221–2
Godin, Louis, 129–30, *130*, 146, 191–2
    see also Peru expedition (Louis Godin's party)
Goethe, Johann Wolfgang von, 102, *104*, 265
Gotland, 221–2
Graham, George, *137*, 138, 141, 195, 233–4, *234*
    instruments built by, 145, 172, 174, *174*, 189, 195, 208, 213–4, 232, *233*, 239
Grand Tour, in general, 99–103, *101*, *104*, 111
Grand Tour, of AC (1732–1736), *22*
    companions, 103, *104*, 119
    England, 133–6, 137–43, 146
    France, 119, 121–5, 127, 131–2
    Italy, 111–9, 281
    journal/notebook, 105, *105*, 110, 127, *127*, 138, 141, 207
    Sweden and Germany, 105–11
gravitation, 126, 239–40
Great Comet (1680), 59, *60*, 240
Great Comet (1744), 251–2, *252*, 253, 256
Greenwich Royal Observatory, 133, *134*, 138, *139*, 141–2, 211

Greifswald, 54–5, 106–7, *106*
Guen, Adam, 150, 152
Gunillaklockan tower and bell, Uppsala, 41, *42*
Gustaf II Adolf, King of Sweden, 49, 52
Gustavianum, Uppsala University, 39, *40*, *42*, 246
Gyllengborg, Carl, 210, 255, 256

Halley, Sir Edmond, 138, 139, *140*, 141, 241, 242
Hallgren, Christoffer, 272
Hårleman, Carl, 210–1, *211*, *212*, 256, 265
Hellant, Anders, 167, 175
Helsingør, 154, *154*
Herbelot, M., 149, *176*, 194
Hiärne, Urban, 218–9
Hilster, Nicolàs de, 231–2
Hiorter, Olof, 209, *209*, 234–5, 251–2, 254, 255, 256, 259, 260
Höpken, Baron Anders Johan Von, 68, 253, *258*

ice ages, 226–7, 229–30, 277
  post-glacial uplift, 228–30
Iggön Island, *20*, 219–20, *220*, 221
International Bureau of Weights and Measures (BIPM), 248–9, *249*

Jamaica, 136–7
Jamieson, Thomas, 227–8, *228*
Jupiter and moons, 108, 141, 260–1

Kallioinen, Miia, 161
Kelvin scale and absolute zero, 247–8, 249
King's Barn project, 71, 89
Kirch, Christfried, 107–9, *108*
Kirch, Christine and Margaretha, 108
Kirch, Maria-Margarethe, 108, *108*, *110*, 241
Klingenstierna, Samuel, 70–1, *70*, 89, 90, 91–2, 121, 208, 241, 255
Körner, Johan Christopher, 211
Korteniemi, Tuomo, 167, 175, *179*
Korteniemi, Veli-Markku, 167–8, 175
Korteniemi guesthouse, 174–5, *176*

La Condamine, Marie, 129–30, 192, 202, *203*
  see also Peru expedition (Louis Godin's party)
Lagergren, Karl, 250
Lapland
  Anders Spole's expedition, 60–1, *63*, 68–9, *69*, 95
  Carl Linnaeus' expedition, 95–8, *96*, *97*

Pierre de Maupertuis' expedition *see* Arctic expedition (1736–1737)
latitude distance measurement
  Arctic expedition *see under* Arctic expedition (1736–1737)
  Peru expedition, 129, 159, 161, 192–3, 202, *203*
latitude location measurement, 91
Le Monnier, Pierre Charles, 131, 184, 187, 197
Leipzig, *22*, 109
Lenngren, Anna Maria, 52
light, speed of, 113, 140
Lindbom, Anna-Zara, 246
Linnaeus, Carl, 92–8, *94*, *96*, *97*, 221, 224–5, 242, 246, 264–5, *266*–7
Locke, John, 66
London, *22*, 127, 133–40, *134*, *135*, 141–5
longitude measurements, 260–1
Louis XV, King of France, *128*, 130, 183, 241
Lövgrund Island, *20*, 215–6, *219*, 225
Luleå, 218
Lund University, *20*, 55–6, 223

Machin, John, 143, 195
magnetic field, of Earth, 232–6
Mairan, Jean-Jacques d'Ortous de, 199
Malthus, Thomas, 262–3, *263*
Manfredi, Eustachio, 112–3, *114*, 117
Manfredi, Maddalena and Teresa, 113
Maupertuis, Marie de, 193–4
Maupertuis, Pierre Louis Moreau de, 127–8, 130–2, 185–9, 190, 193–4, 195–9
  Foundation in name of, 167
  friendship with AC, 183, 186, 196, 254
  portraits, *128*, 197, *198*
  publications, 195, *195*, 199–200, *200*
  quoted, 145–6, 170–1, 174, 178, 186, 187–8, 189, 193–4, 196, 197, 199, 201
  see also Arctic expedition (1736–1737)
Maupertuis monument, Pello, 161–3, *162*, 183
Maurepas, Comte de, 149, 183, 186–7, 199
Meldercreutz, Jonas, 71, 103, *104*, 134, 159
  see also Grand Tour, of AC (1732–1736)
meridians, 112, 113, 115, 124, 125, 141–2, *142*
  measurements, 125, 126, 164, 195–7
  see also latitude distance measurement
mining, *20*, 83–8, *84*, *85*, 278
Montague, Lady Mary Wortley, 145
Moon, 89, 263
Moraea, Sara Lisa, 224–5, 264
Mortimer, Cromwell, 135–6, 143, 188
Mount Horrilakero, 171
Mount Huitaperi, 169

Index

Mount Kakamavaara, 168, 169
Mount Kittisvaara, 161–3, 170, 172, 174–6, *176*, 181, 183
Mount Niemivaara, 168, 170, *170*
Mount Paderno, 115–6
music, and mathematics, 75–6, *77*
mythology, 29–31

National Board of Antiquities, 52
Newton, Sir Isaac, *125*, 126, 180, 181
Nils, Rik, 219, 220, *220*
Northern Lights, 32, 108, 109–11, *110*, 129, 141, 152, 234–5
Northern Wars, 56
Nuremberg, *22*, 109–11

observation, measurement and scientific method
    AC's appreciation of importance, 32, 35
    and the Enlightenment, 66
    importance for climate change understanding, 34
    Johannes Bilberg on, 69
    Nils Celsius' clash with the Church, *58*, 59, 66
    see also weather observations
observatories
    Berlin, 107–9, *107*
    Bologna, 111–3, *114*
    Castricum, 231–2
    Greenwich, 133, *134*, 138, *139*, 141–2, 211
    Mount Kittisvaara, 174, *176*, 181
    Paris, *123*, 124–5, 187, 211
    Stockholm, 256, 260, *260*
    Tornio, 175–6, 181
    Uppsala, 11–2, 14, *21*, 54, 59, 208, 210–3, *212*, 214, 250, 251–2, 255–6, 259
Östhammar and Öregrund, 217
Outhier, Abbé Réginald, 131, 149–50, 152, 154, 156, 157, 158, 159, 166

Padua, 111, 241
Paris, *22*, 121–4, *123*, 193–4, 196–7
    l'Académie des Sciences, 124–5, 129, 181, 187–9, *189*, 190, 197, 201
    International Bureau of Weights and Measures (BIPM), 248–9, *249*
    Versailles, 186–7, *187*, 241
Pascal, Blaise, 81–2, 83
Pello, *20*, 161–3, *162*, 175
Peru expedition (Louis Godin's party), 129–30, 146, 159, 191–3, *191*, *192*, *193*, 202
Picard, Jean, 125–6, 196, 199

planetary observations, 108, 141, 260–1
Planström, Christina, 180, 184, 193–4
Planström, Elisabeth, 180, 184, 193–5
population growth, 262–3
Pound, James, 141
Le Prudent (ship), 150–4, *151*, 155, 156, 159, 163, 184
    see also Arctic expedition (1736–1737)
pulka sledges, 181, *182*
Pythagoras, 75–6, *77*

Réaumur, René Antoine Ferchault de, 188, 242, 243, *244*
Reformation, Sweden, 47–9
Rome, *22*, 116–8
Rømer, Ole Christensen, 241, 242, *244*
Rowlandson, Thomas, 122, *123*
Royal Society, London, 88, 142–3, *144*, 188, 199
Royal Society of Sciences, Sweden, 63, 88, 89, 181, 208–9, *247*
Royal Swedish Academy of Sciences, 222–3, 246, 256, 260
Rudbeck, Olaf (the Elder), 31, *31*, *38*, 39, 54, 92, 181, 280
Rudbeck, Olaf (the Younger), 61, *62*, *63*, 93, 95
Rudman, Johan, 219
Runby, *216*, 217
Runeby, Jörgen, 250

Sala silver mine, *20*, 83–7, *84*, *85*
Scheffel, Johan Henrik, 250, 257–8
Schimper, Karl Friedrich, 227
Scotland, raised beaches, 227–8, *228*
sea level changes, 29, 32, 215–24, *216*, *219*, *220*, 225–30, *228*
'seal rocks,' 215–6, 219–20, *219*, *220*, 225
Seniergues, Jean, 192, *192*
ship burials, 29–31, *30*, 32–3, *32*
Sixtus VI (Pope), 46, *46*, 47
Sloane, Sir Hans, 136–8, *137*, 143, 171, 264
Sommereux, M., 149
Spole, Anders, 39, 54–7, 58–64, *58*, 68, 95, 103, 214
Spole, Anna Maria, 64
Spole, Gunilla, 37–9, 55, 60, 69, 134, 253, 254, 257–9
    dining house, 67–8, 157, 208, 232–3
    letters from AC to, 113, 132
Stärkman, Petra, 250
Stempels, Eric, 213–4
Stevens, Tom, 271, 272, 275, 277
Stockholm, *22*, *153*, 184, 225–6, 256, 260, *260*, 261–2

Stockholm Sea Level Series, 225–6
Stone of Käymäjärvi, 181–3, *182*
Strömer, Mårten, 70–1, *254*, 255, 256
Sun, 231–2, 233–6, 261, 274
sustainability, 271–2, 278–80, *279*

thermometry and temperature scales, 78, 240–9, *241*
Thomson, William, 1st Baron Kelvin, 247, *247*
time, deep (past and future), 271, 274
Tolvanen, Janne, 161
tombstone engraving, Mount Paderno, 115–6, 281
Torne, River, 177–9, *179*
Tornio, *20*, 159, *160*, 163–4, *164*, 166–8, 171, 172–4, *173*, 175–6, 180–1
Torricelli, Evangelista, 81, 83, 86
Tournières, Robert Le Vrac de, 197, *198*
tuberculosis *see* consumption (tuberculosis)

Ulfsson, Jacob, 217
Ulfvebrand, Niklas, 83, 84, 86
Ulrica Eleonora, Queen of Sweden, 155, *156*, 184
Ulvsson, Jakob, Archbishop of Uppsala, 46, *46*, 47
universe, fate of, 274
Uppland region, 29, 32, 216–7
  see also Gamla Uppsala; Uppsala
Uppsala, *20*, *21*
  Arctic expedition visit to, 157–8
  botanical garden, 92, 265
  Carolina Rediviva Library, 73
  castle, 39, 41, *42*
  Great Fire (1702) and reconstruction, 37–41, *38*, 95, 280
  Gunillaklockan tower and bell, 41, *42*
  Gunilla's dining house and family home, 67–8, 157, 208, 232–3
  observatories, 11–2, 14, *21*, 54, 59, 208, 210–3, *212*, 214, 250, 251–2, 255–6, 259
  present day, 11–2, 14, 73, 250, 256
  statue of AC, 12, 240, 256

Uppsala Cathedral, 39, *40*, 41, *42*, *46*, 64, 75, 82–3, *82*, 92, 265, 280
Uppsala University
  AC's unpaid appointments, 89, 90
  Anders Spole and, 57–9
  Ångströmlaboratoriet, 213–4, 254
  astronomy professorship, 49, 52, 57, 63, 75, 253
    AC's professorship, 91, 121, 132, 207, 208–10
  detention house, 52
  funds observatory, 210, 211, 213
  Geocentrum, 271, 272
  Gustavianum, 39, *40*, *42*, 246
  history of, 46–9, *46*
  history professorship, 54, 265
  mathematics professorships, 70, *104*
  medicine professorship, 264–5
  permaculture garden, 271–2
  rectorship, 252, 265
Uppsala Weather Series, 78, 81, 91, 240, *240*, 243, 246–7, 272, *273*
Urtel, Lena, 250

Venetz, Ignaz, 226–7
Venice, *22*, 111
Versailles, 186–7, *187*, 241
Vikings, *216*, 217
Voltaire, 128, 194, 197, 199

Wargentin, Pehr, 70–1, 209, 246, 256, 259–62, 263
weather observations
  AC leads on, 91, 240, *240*, 243, 246–7
  Eric Burman and AC, 78, 81, *273*
  Maria-Margarethe Kirch, 108
  Pehr Wargentin, 261–2
  Uppsala Series, 78, 81, 91, 240, *240*, 243, 246–7, 272, *273*
  see also barometers and air pressure measurement; thermometry and temperature scales
Wren, Sir Christopher, 133, 135, 138

'Zeno' paradox, 72